WATER MOVEMENT THROUGH LIPID BILAYERS, PORES, AND PLASMA MEMBRANES
Theory and Reality

*Distinguished Lecture Series of the
Society of General Physiologists
Volume 4*

Distinguished Lecture Series of the Society of General Physiologists

Water Movement Through Lipid Bilayers, Pores, and Plasma Membranes
Theory and Reality

Distinguished Lecture Series of the
Society of General Physiologists
Volume 4

Alan Finkelstein

Departments of Physiology & Biophysics and Neuroscience
Albert Einstein College of Medicine
Bronx, New York

A Wiley-Interscience Publication
JOHN WILEY & SONS

New York · Chichester · Brisbane · Toronto · Singapore

Library of Congress Cataloging in Publication Data:

Finkelstein, Alan.
    Water movement through lipid bilayers, pores, and plasma membranes.

    (Distinguished lecture series of the Society of General Physiologists; v. 4)
    "A Wiley-Interscience publication."
    Bibliography: p.
    Includes index.
    1. Plasma membranes. 2. Bilayer lipid membranes.
3. Water. 4. Biological transport. I. Title.
II. Series. [DNLM: 1. Biological Transport. 2. Body Fluids—metabolism. 3. Cell Membrane Permeability.
W1 DI842S v.4 / QU 105 F499w]

QH601.F57   1987   574.87'5   86-18973
ISBN 0-471-84787-9

Printed in the United States of America

10 9 8 7 6 5 4 3 2 1

To Alex Mauro, my friend, colleague, and mentor, who first introduced me to the subtleties and complexities of water movement through membranes and whose incessant interest and enthusiasm for the subject have since been a continuing provocation and stimulus to me.

Preface

The movement of water across cell membranes has long been a central topic in general physiology. It is also a topic that from its inception has wedded, often in an uneasy marriage, biological and physical scientists. Thus, the early experimental investigations of osmosis by the plant physiologist Pfeffer (1877) stimulated the physical chemist van't Hoff (1887) to provide a physical theory for osmotic equilibrium; efforts by physiologists, physicists, and physical chemists since then to provide equally satisfactory theories for nonequilibrium situations, that is, for osmotic flow of solvent across membranes, continue to the present day. With the advent of readily obtainable isotopically labeled water in the late 1940s, another measure of water transport across membranes became available, and with it additional questions to confound experimentalists and theoreticians. The formalism of irreversible thermodynamics was also added to the impedimenta of transport savants about 30 years ago (Kedem and Katchalsky, 1958), and its contributions to the subject of solvent and solute movement across membranes have remained a mixed blessing ever since.

Physiologists' interests in the theory of water transport have not been exclusively, or even primarily, motivated by a desire to plumb the depths of physicochemical theory. A major stimulus has been the conviction that the proper application of theory to water permeability measurements can address important questions concerning the route of water (and solute movement) through cell membranes. For example, is the main pathway of water movement into and out of cells through the bilayer proper of the plasma membrane or through aqueous pores in that membrane? If it is through pores, what are their number and size? Is water transport across epithelia primarily transcellular or intercellular? The pursuit of answers to such questions has directed physiologists' attention to permeability measurements on artificial phospholipid bilayer membranes and to the insights that they provide. Conclusions reached from studies on these membranes have had an important influence both on the interpretation of biological data and in the molding of ideas about transport across cell membranes.

This monograph on the water permeability of lipid bilayers and cell membranes is divided into three sections: Part I presents the theoretical framework for treating water movement across membranes; Part II reviews the results of water permeability measurements on artificial lipid bilayer membranes in the context of this framework; Part III discusses permeability data on biological membranes in the light of the material presented in the preceeding two parts. None of these sections, particularly the last, is exhaustive in its treatment of its subject matter. The topics treated in Part I, and the nature of their treatment, reflect my view of those issues that I feel are essential to a physiologist's understanding of water transport across cell membranes. A possible important omission from that section is the topic of electroosmosis, which is considered only in the context of single-file transport. The biological systems selected as examples in Part III are ones with which I am most familiar; hopefully they are of general interest and the principles that they illustrate have wide applicability. In the ordering and presentation of material, I have had in mind both the general reader—physiologist, biophysicist, physical chemist, and students in these disciplines—who wishes to become acquainted with the subject of water transport across cell membranes, and the specialist actively engaged in research in this area. I have strived for an approach that is neither too technical and obscure for the former, nor too pedantic and trivial for the latter.

ALAN FINKELSTEIN

Contents

CONTENTS

CONTENTS

CONTENTS

PART

I

THEORY

Since our interest is a biological one, we take water as the solvent throughout this section; this causes no loss in generality. We also confine our attention, for the most part, to nonelectrolyte solutions. The formidable and very interesting problems attendant to the treatment of electrolytes are beyond the scope of this work, although occasional reference is made to some of them. We further restrict our attention to the dilute solution limit, where activity coefficients are set equal to one.

1

Osmotic Equilibrium

Consider a membrane separating two compartments; one of them (compartment 2) contains pure solvent, water, and the other (compartment 1) contains a solution of solute s (Fig. 1-1). The membrane is assumed to be permeable to the solvent but *absolutely impermeable to the solute.* We make no specification of the physicochemical mechanism by which the membrane permits solvent to cross but prevents solute passage from one compartment to the other. The reader may invoke any fantasy that he or she chooses to account for the membrane's permselectivity. For example, the membrane may contain pores wide enough to admit water molecules but too narrow to allow solute entry; alternatively, it may consist of a material in which water can dissolve, and thereby traverse the membrane, but in which the solute is totally insoluble. Later, in Chapter 2, when we consider solvent transport (i.e., osmotic flow or osmosis), we shall have to be concerned with the nature of the membrane and the mechanism of

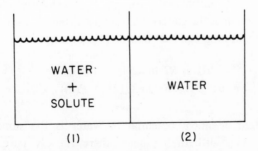

Figure 1-1. A membrane separating a solution of an impermeant solute from pure solvent.

solute impermeability. The analysis given below of the equilibrium situation, however, is a thermodynamic one and is therefore completely independent of mechanistic considerations.

There are a number of alternative standard textbook ways of treating osmotic equilibrium. The one used here invokes the general Gibbsian equilibrium condition that the chemical potential μ_i of species i is the same at every point in the system to which it has access. In the present instance, this means that the chemical potential of water μ_w is the same in compartments 1 and 2; that is,

$$\mu_w(1) = \mu_w(2) \tag{1-1}$$

Note that a similar condition does not apply to the solute, because, by definition, the solute does not have access to both compartments. (This remains true even if solute s is present in both compartments, since it is not free to move between them; thus, s in compartment 1 does not have access to compartment 2, and similarly, s in compartment 2 does not have access to compartment 1.) For the dilute solution case being considered, the chemical potential of water is given by

$$\mu_w = \mu_w^{(o)} + RT \ln X_w + P\overline{V}_w \tag{1-2}$$

where $\mu_w^{(o)}$ is the standard chemical potential of water, X_w is its mole fraction, $\overline{V}_w$ is its partial molar volume, P is the hydrostatic pressure of the solution, R is the gas constant, and T is temperature in degrees Kelvin. Substituting eq.(1-2) into eq.(1-1) we have

$$\begin{aligned} &\mu_w^{(o)}(1) + RT \ln X_w(1) + P(1)\overline{V}_w \\ = \; &\mu_w^{(o)}(2) + RT \ln X_w(2) + P(2)\overline{V}_w \end{aligned} \tag{1-3}$$

The standard chemical potential of water is the same in both compartments. Furthermore, since there is only pure water in compartment 2, $X_w(2) = 1$. Equation (1-3) can therefore be rewritten in the form:

$$\Pi \equiv [P(1) - P(2)] = -\frac{RT}{V_w} \ln X_w(1) \qquad (1\text{-}4)$$

The quantity Π is called the osmotic pressure of solution 1. It is the hydrostatic pressure difference that must exist between compartments 1 and 2 in order for there to be no net water movement across the membrane. At this pressure difference the chemical potential of water is the same in both compartments and the system is at equilibrium. The magnitude of Π is given by eq.(1-4), and we see that since $X_w(1) < 1$, Π is positive; that is, the hydrostatic pressure of the solution phase (s in water) is greater than that of the pure solvent phase. (We shall discuss shortly how this pressure difference can arise or develop experimentally.)

Equation (1-4) is often rewritten in terms of the solute s rather than in its present form in terms of the solvent, water. Since,

$$X_w + X_s = 1$$

and, for dilute solutions, $X_s \ll 1$, eq.(1-4) becomes:

$$\Pi \approx \frac{RT}{V_w} X_s$$

By definition:

$$X_s \equiv \frac{n_s}{n_w + n_s}$$

where n_s is the number of moles of solute and n_w is the number of moles of water. For dilute solutions, $n_s \ll n_w$; therefore,

$$X_s \approx \frac{n_s}{n_w}$$

Substituting this into the above expression for Π we get,

5

$$\Pi \approx RT\frac{n_s}{\overline{V}_w \mathbf{n}_w} = RT\frac{n_s}{V_w}$$

where V_w is the total volume of water. But for dilute solutions, the total volume of water is essentially the volume of the solution. Thus, in the dilute solution limit we finally obtain:

$$\Pi = RTc_s \tag{1-4a}$$

where c_s is the molar concentration of solute in compartment 1. If we had started with solute in both compartments, the same derivation would lead to:

$$\Delta\Pi = RT\,\Delta c_s \tag{1-4b}$$

where

$$\Delta c_s \equiv c_s(1) - c_s(2) \tag{1-5a}$$

is the difference in the solute concentration in the two compartments, and

$$\Delta\Pi \equiv \Pi(1) - \Pi(2) \tag{1-5b}$$

Equation (1-4a) [or (1-4b)] is the famous van't Hoff expression for osmotic pressure. It is the pressure difference that must exist at equilibrium across a semipermeable membrane separating dilute, ideal solutions. [For more concentrated, nonideal solutions, an activity coefficient for water can be introduced into eq.(1-4); alternatively, eq.(1-4a) can be expressed in the form

$$\Pi = RTc_s + Bc_s^2 + Cc_s^3 + \cdot\cdot\cdot$$

where $B, C, \ldots$ are the so-called virial coefficients for the solute (see, for example, Tanford, 1961).] It is instructive to consider two of

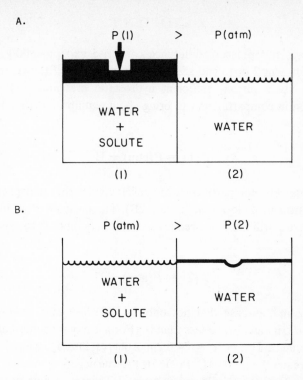

Figure 1-2. Two ways of experimentally establishing the osmotic pressure difference Π across a membrane separating a solution of an impermeant solute from pure solvent. (A) The pure solvent in compartment 2 is open to the atmosphere at pressure $P(\text{atm})$, and a pressure $P(1)$, sufficient to stop solvent flow, is applied via a piston to the solution in compartment 1. $P(1) = P(\text{atm}) + \Pi$. (B) The solution in compartment 1 is open to the atmosphere at pressure $P(\text{atm})$; compartment 2 is filled with solvent and closed off. Because of the minute amount of water that has left compartment 2 (by flowing, across the membrane, into compartment 1), the pressure there has fallen below atmospheric pressure to $P(2)$, at which value solvent flow is stopped. (The subatmospheric pressure in compartment 2 is indicated by the indentation of a pressure transducer located in the lid of the compartment.) $P(2) = P(\text{atm}) - \Pi$. Note that if Π is greater than one atmosphere (that is, the concentration of solute in compartment 1 is greater than 0.04 M), $P(2)$ is negative, and the solvent in compartment 2 is under tension.

7

the ways this pressure difference can arise experimentally. With compartment 2 open and at atmospheric pressure $P(\text{atm})$, one can apply, via a piston, sufficient hydrostatic pressure $P(1)$ to the solution in compartment 1 to bring about equilibrium (Fig. 1-2A); that is,

$$P(1) = P(\text{atm}) + \Pi \qquad (1\text{-}6a)$$

Alternatively, compartment 1 can be left open to the atmosphere and compartment 2 closed off (Fig. 1-2B). The pressure $P(2)$ in compartment 2 will fall to the required level for equilibrium to exist; that is,

$$P(2) = P(\text{atm}) - \Pi \qquad (1\text{-}6b)$$

Notice in this case that not only is $P(2)$ less than atmospheric pressure, it can also be less than 0. [For a 1 molar solution at room temperature, $\Pi \approx RTc_s \approx 24.6$ atmospheres; thus, from eq.(1-6b), if c_s is greater than $0.04\ M$, $P(2) < 0$.] A true negative pressure (i.e., a tension state) can exist in compartment 2. There is nothing mysterious or metaphysical about negative pressures, or tension states, in liquids. A simple way to achieve such a state is to completely fill a syringe with water, close off the outlet, and pull back on the plunger with sufficient force (Fig. 1-3). By this procedure a negative pressure can be applied

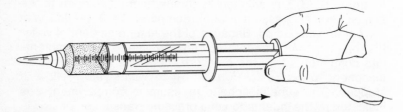

Figure 1-3. One way of putting water in a tension state, that is , placing it under negative pressure. A syringe, completely filled with water, is closed off at the outlet, and the plunger is pulled back with sufficient force.

to the water in the syringe. Of course, if nucleation occurs and a bubble of water vapor forms, the plunger separates from the fluid and the pressure within the syringe suddenly rises to a positive value that eventually equals the vapor pressure of water. Negative pressures have been achieved in liquids by several different means (see Hayward, 1971), including the osmotic equilibrium method depicted in Figure 1-2B (Mauro, 1965; 1981).

2

Osmotic Transport (Osmosis) Induced by an Impermeant Solute

Let us return to the situation depicted in Figure 1-1. We have just seen that if ΔP [i.e., $P(1) - P(2)$] is equal to Π, the system is at equilibrium, and there is no flow of water across the membrane (Figs. 1-2A and 1-2B). If $\Delta P > \Pi$, water flows across the membrane from compartment 1 to compartment 2, whereas if $\Delta P < \Pi$, it will flow in the opposite direction (Fig. 2-1). Assuming a linear pressure-flow relation (which must be valid at least for small deviations from equilibrium), we can write

$$J_v = L_p(\Delta P - \Pi) \qquad (2\text{-}1a)$$

or, more generally, if impermeant solute is present in both compartments

$$J_v = L_p(\Delta P - \Delta\Pi) \qquad (2\text{-}1b)$$

where J_v is the volume flux (e.g., in cm^3/s) across the membrane, and L_p is the hydraulic permeability coefficient of the membrane. It is sometimes convenient to rewrite eq.(2-1b) in the form:

$$\Phi_w = \frac{J_v}{\overline{V}_w} = P_f A \left(\frac{\Delta P}{RT} - \Delta c_s\right) \qquad (2\text{-}2)$$

where Φ_w is the flux of water in moles/s, A is the membrane area, and P_f is the osmotic, or filtration, permeability coefficient (in units of cm/s) and is related to L_p through the equation:

10

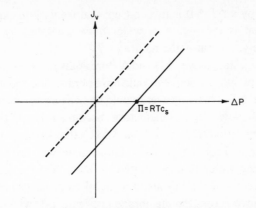

Figure 2-1. A sketch showing the equivalence of hydrostatic pressure (ΔP) and osmotic pressure (Π) in causing solvent to flow across a membrane. The dashed line is a plot of solvent flow (J_v) vs. ΔP for a membrane with pure solvent present on both sides. (The J_v–ΔP relation is assumed to be linear over the range of values being explored.) The solid line is a similar plot for the same membrane separating a solution of impermeant solute (side 1) from pure solvent (side 2), as in Figure 1-1. Note that there is simply a parallel displacement of the original plot (if c_s is not too large), and that instead of flow being zero when $\Delta P = 0$, it is now zero when $\Delta P = \Pi = RTc_s$; at this pressure there is osmotic equilibrium. Note further that when $\Delta P = 0$, there is flow (osmosis) from compartment 2 to compartment 1 that is equivalent to that which would have been produced, in the absence of solute, by a ΔP equal to $-\Pi$. Thus, ΔP and Π (or $\Delta \Pi$, if there is solute in both compartments) are equivalent in their ability to produce solvent flow across a membrane.

$$P_f = \frac{L_p RT}{\overline{V}_w A} \qquad (2\text{-}3)$$

Equation (2-1b), an experimentally verifiable relation, expresses the equivalence between ΔP and $\Delta \Pi$ in driving water across a membrane (see Fig. 2-1). Implicit in that equivalence is the concept that the mechanism of fluid transport generated by a $\Delta \Pi$ is the same as that generated by a ΔP.* It is this concept that we wish to examine. We ask, "What is the nature of water flow generated by $\Delta \Pi$?"; or alternatively, "What is the nature of P_f?"

If as suggested above, $\Delta \Pi$ and ΔP are equivalent and generate the same type of fluid motion, we could have phrased the above questions in terms of ΔP. This, however, would be begging the issue. It would also be pedagogically unsatisfactory, because it is just this equivalence of flows produced by ΔP's and $\Delta \Pi$'s that students and investigators often find confusing. Many, for instance, who feel comfortable in discussing water flow generated by a hydrostatic pressure difference across a membrane are brought up short if asked what causes water to move across the membrane in Figure 1-1, when both compartments are open to the atmosphere and hence there is no ΔP across the membrane. It is precisely this question that we wish to address. In contrast to the equilibrium situation, however, in which the mechanism of solute exclusion by the membrane is irrelevant to the considerations, osmotic water flow (or, for that matter, transport of any substance across a membrane) cannot be discussed without some specification of membrane composition. In what follows, we consider two types of membranes: (1) lipoidal or "oil" membranes and (2) porous membranes. We focus on these two types because of their relevance to our discussions in Parts II and III of lipid bilayers and cell membranes.

A. "OIL" MEMBRANE

By an oil membrane we mean one that is composed of a material (e.g., oil or wax) in which water is very poorly soluble. Water crosses the membrane by a solubility-diffusion mechanism. That is, it partitions into the membrane at each membrane-solution interface, and moves through the membrane from one interface to the other by simple diffusion. Since water is very poorly soluble in the membrane phase (i.e., its partition coefficient into the membrane phase is very

---

*In Chapter 5, we note a subtle difference in the fluid velocity profile within pores for flow generated by a ΔP and flow generated by a Δc_s of a *permeant* solute.

small), its concentration in the membrane is very low, and hence inidividual water molecules move independently of one another; they do not "see" each other in the membrane.

Let us consider osmosis of water across a membrane separating solutions 1 and 2 (at the same hydrostatic pressure P) that contain different concentrations of impermeant solute. From the above description of an oil membrane, we can directly derive an expression for P_f in terms of the partition coefficient (K_w) of water into the membrane and its diffusion coefficient (D_w) within that phase (Fig. 2-2). Assume that the interfacial kinetics are fast compared to diffusion

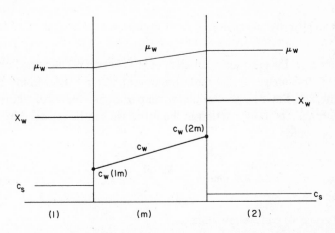

Figure 2-2. Profiles of the concentration (c_w) and chemical potential (μ_w) of water in an "oil" membrane separating solutions of unequal concentrations (c_s) of impermeant solute. "Just within" the membrane at the two interfaces, water is in partition equilibrium with the adjacent solution, and this fixes the concentrations there at $c_w(1m)$ and $c_w(2m)$. [$c_w(1m)$ is less than $c_w(2m)$ because the mole fraction (X_w) of water is lower on side 1, as a consequence of the higher solute concentration there.] The flux of water between the two interfaces (and hence its flux across the membrane) is by simple diffusion. Note that although there is a discontinuous change in water concentration at the interfaces, the chemical potential of water is continuous across the interfaces.

through the membrane, which can always be realized if the membrane thickness is large enough. Then essentially equilibrium exists at the boundaries, and therefore the value of the chemical potential of water in the aqueous phase, μ_w (water phase), equals that "just inside" the membrane, μ_w (oil phase):

$$\mu_w^{(o)}(\text{water phase}) + RT \ln X_w(\text{water phase}) + P\bar{V}_w(\text{water phase}) =$$
$$\mu_w^{(o)}(\text{oil phase}) + RT \ln X_w(\text{oil phase}) + P\bar{V}_w(\text{oil phase})$$

$$\frac{X_w(\text{oil phase})}{X_w(\text{water phase})} \equiv K_w$$

$$= \exp[(\mu_w^{(o)}(\text{water phase}) - \mu_w^{(o)}(\text{oil phase}) + P\,\Delta\bar{V}_w)/RT] \quad (2\text{-}4)$$

where K_w is the partition coefficient of H_2O between water and "oil," and $X_w(\text{oil phase})$ is the mole fraction of water "just within" the membrane. Since by assumption water partitions very poorly into the membrane, its mole fraction in the oil phase can be replaced by its concentration there:

$$c_w(\text{oil phase}) = \frac{X_w(\text{oil phase})}{\bar{V}_{\text{oil}}} \quad (2\text{-}5)$$

and eq.(2-4) can be rewritten as:

$$c_w(\text{oil phase}) = K'_w X_w(\text{water phase}) \quad (2\text{-}6)$$

where $$K'_w = \frac{K_w}{\bar{V}_{\text{oil}}} \quad (2\text{-}7)$$

[Note that although eq.(2-6) was derived under the assumption that interfacial kinetics are fast compared to diffusion within the membrane, it remains valid even without that assumption, except that K'_w is given by:

$$K'_w = \frac{\alpha K_w}{\overline{V}_{\text{oil}}} \tag{2-8}$$

where $0 < \alpha < 1$.] For dilute solutions eq.(2-6) can be rewritten in terms of the solute concentration c_s:

$$c_w(\text{oil phase}) = K'_w(1 - \overline{V}_w c_s) \tag{2-9}$$

Equation (2-9) applies at each interface to the concentration of water "just within" the membrane. The flux Φ_w of water between these interfaces (and hence the osmotic flow across the membrane) occurs by simple diffusion, and from Fick's law is, for a membrane of area A and thickness δ,

$$\Phi_w = \frac{D_w A}{\delta} \Delta c_w(\text{oil phase}) \tag{2-10}$$

where D_w is the diffusion constant of water within the membrane phase, and

$$\Delta c_w(\text{oil phase}) = c_w^{1m}(\text{oil phase}) - c_w^{2m}(\text{oil phase})$$

$$= -\frac{K_w \overline{V}_w}{\overline{V}_{\text{oil}}} \Delta c_s \tag{2-11}$$

where $c_w^{1m}(\text{oil phase})$ and $c_w^{2m}(\text{oil phase})$, obtained from eq.(2-9), are the concentrations of water "just within" the membrane at interfaces 1 and 2, respectively (see Fig. 2-2). Substituting eq.(2-11) into eq.(2-10) we obtain

$$\Phi_w = -P_f A \Delta c_s = -P_f A \frac{\Delta\Pi}{RT} \quad \text{(in the absence of a } \Delta P) \tag{2-12}$$

15

CHAPTER 2

where,

$$P_f = \frac{D_w K_w \overline{V}_w}{\delta \, \overline{V}_{oil}} \qquad \text{(for an oil membrane)} \qquad (2\text{-}13)$$

Osmosis occurs by a solubility-diffusion mechanism; the osmotic permeability coefficient P_f is directly proportional to the partition coefficient (K_w) of water between the membrane phase and aqueous solution, and to the diffusion constant (D_w) of water within the membrane phase.

One can also explicitly demonstrate for an oil membrane the equivalence of water flow generated by a hydrostatic pressure difference ΔP to that generated by an osmotic pressure difference $RT \, \Delta c_s$. Consider an oil membrane separating two compartments, 1 and 2, of pure water at hydrostatic pressures P_1 and P_2 respectively, with $P_1 > P_2$. For an oil membrane this requires a mechanical support, such as the cell wall support for the plasma membrane of plant cells. Within the membrane the pressure is constant at the higher value P_1 (Fig. 2-3). Again assuming that equilibrium conditions prevail at the interfaces, that is, the chemical potential of water is continuous across each interface, we can write for interface 1, from eq.(2-9) with $c_s = 0$:

$$c_w^{1m}(\text{oil phase}) = K_w' = \frac{K_w}{\overline{V}_{oil}} \qquad (2\text{-}9a)$$

For interface 2 we have:

$$\mu_w^{(o)}(\text{water phase}) + RT \ln X_w(\text{water phase}) + P_2 \overline{V}_w(\text{water phase}) =$$
$$\mu_w^{(o)}(\text{oil phase}) + RT \ln X_w(\text{oil phase}) + P_1 \overline{V}_w(\text{oil phase}) \qquad (2\text{-}4a)$$

From the definition of K_w [eq.(2-4)] and given that X_w(water phase) $= 1$, eq.(2-4a) can be rearranged to give:

16

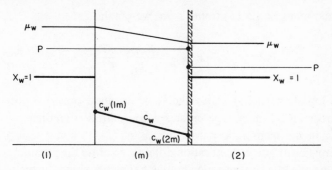

Figure 2-3. Pressure, concentration, and chemical potential profiles within an "oil" membrane separating compartments of pure water at different hydrostatic pressures [$P(1) > P(2)$]. The oil membrane is given mechanical support at the right interface by a porous network, freely permeable to water, that does not allow the "oil" to leak through. (This is analogous to the cell wall support for the plasma membrane of plant cells.) $c_w(2m)$ is lower than $c_w(1m)$ because of the difference in pressure across interface 2; the lower value of $c_w(2m)$ compensates for the higher pressure within the membrane, so that μ_w is continuous across the interface. Note that there is no pressure gradient within the membrane; the hydrostatic pressure difference between the two solutions is converted into a concentration gradient of water within the membrane.

$$X_w^{2m}(\text{oil phase}) = K_w \exp[-\Delta P \overline{V}_w(\text{water phase})/RT] \quad (2\text{-}14)$$

where,

$$\Delta P \equiv P_1 - P_2$$

Confining ourselves to small ΔP's, we obtain upon substituting eq.(2-5) into eq.(2-14):

$$c_w^{2m}(\text{oil phase}) = \frac{K_w}{\overline{V}_{\text{oil}}} \left[1 - \frac{\Delta P \overline{V}_w}{RT} \right] \quad (2\text{-}15)$$

17

Subtracting eq.(2-15) from (2-9a) we obtain:

$$\Delta c_w(\text{oil phase}) = \frac{K_w \overline{V}_w}{\overline{V}_{\text{oil}}} \frac{\Delta P}{RT} \tag{2-16}$$

Equation (2-16) states that a hydrostatic pressure difference across an oil membrane creates a concentration gradient of water within the membrane phase, just as eq.(2-11) states that an impermeant solute concentration difference across the membrane also gives rise to a concentration gradient of water within the membrane. Thus, the mechanism of water flow generated by a Δc_s or a ΔP across an oil membrane is the same: namely, diffusive flow as described by the integrated form of Fick's law in eq.(2-10). In fact, if we substitute eq.(2-16) into eq.(2-10) we obtain:

$$\Phi_w = P_f A \frac{\Delta P}{RT} \qquad \text{(in the absence of a } \Delta c_s) \tag{2-17}$$

and comparing this with eq.(2-12) we see that the same permeability coefficient P_f applies both to water flow generated by an osmotic pressure difference ($\Delta \Pi$) and to water flow generated by a hydrostatic pressure difference (ΔP). This, of course, is observed experimentally independent of the nature of the membrane, but has been explicitly demonstrated in this instance for an oil membrane. I chose, for pedagogical purposes, to derive separate expressions for water flow in the presence of either a Δc_s alone [eq.(2-12)] or a ΔP alone [eq.(2-17)]. It is easy to show that with both present, the resulting expression for the water flux is eq.(2-2)—the algebraic sum of eqs.(2-12) and (2-17). When $\Delta P = RT\Delta c_s$, the system is at equilibrium; there is no concentration gradient of water within the membrane, and water flux (Φ_w) is zero (Fig. 2-4).

B. POROUS MEMBRANE

Consider a membrane consisting of n water-filled pores embedded in an inert, rigid matrix. For simplicity and concreteness assume that

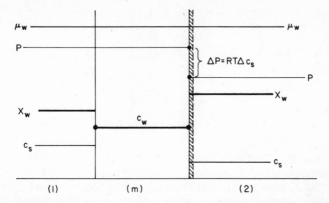

Figure 2-4. Pressure, concentration, and chemical potential profiles within an oil membrane when osmotic equilibrium is established across it ($\Delta P = RT \, \Delta c_s$). (The mechanical support is freely permeable to both water and solute.) There are no gradients of either pressure or water concentration within the membrane; the system is at equilibrium (water flux is zero), and the chemical potential of water is the same in all phases (in the two aqueous solutions and within the oil phase).

the pores are identical, right circular cylinders of radius r and length L. We again turn our attention to Figure 1-1 and ask what causes water to flow, in the absence of a ΔP between the two compartments, when an impermeant solute that cannot enter the pores is present in compartment 1. The question is made particularly piquant if we focus on a small volume element within a pore (Fig. 2-5) and ask what drives the water in that element toward compartment 1. For an oil membrane the answer is simply that water is driven by its concentration gradient (see Fig. 2-2); that is, water diffuses from a higher to a lower concentration region. This clearly is not the answer in the present instance, since with only pure liquid water in the pore, there is no concentration gradient of H_2O. For want of any other explanation, one is forced to conclude that there must exist a pressure gradient (despite the fact the solutions in compartments 1 and 2 are at the same pressure) that acts on any volume element within the pore and drives it

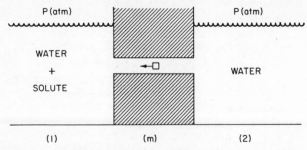

Figure 2-5. A small volume element within a pore of a membrane separating a solution of impermeant solute from pure water. We know from other considerations that osmosis occurs and that therefore water flows through the pore from right to left, but since there is only liquid water in the pore, how does the volume element "know" to move in that direction? That is, what is the force that drives the volume element in that direction?

toward compartment 1. This indeed is the case, and we shall now try to gain some insight into the origin of this pressure gradient within the pore.

Pressure and Pressure Gradient Within the Pore

If, as before, we assume that interfacial kinetics are fast compared to transport between the interfaces, then equilibrium exists at each interface; that is, the chemical potential of water "just within" the membrane (pore) is equal to that in the adjacent compartment. At the interface with compartment 2, which contains pure solvent, this simply means that the pressure "just within" the pore is equal to that in the compartment. At the interface with compartment 1, however, we have:

$$\mu_w^{(o)} + RT \ln X_w(1) + P(1)\bar{V}_w =$$
$$\mu_w^{(o)} + RT \ln X_w(1m) + P(1m)\bar{V}_w$$

where $1m$ refers to the value of quantities "just within" the membrane (pore) at interface 1. But since there is only pure water within the pore,

$X_w(1m) = 1$. For dilute solutions, the above equation then reduces to:

$$P(1) - P(1m) = RTc_s \qquad (2\text{-}18)$$

Equation (2-18) states that there is a fall in pressure (equal to RTc_s) as one enters the pore from solution 1 (Fig. 2-6A). Indeed, this fall in pressure lowers the chemical potential of pure water just within the pore at interface 1 to that of the water in solution 1; the lower pressure of the pure water just within the pore compensates for the lower mole fraction of water in solution 1, thereby equalizing the chemical potential of water on the two sides of the interface. Consequently, since $P(1m) < P(2m)$, there exists a pressure gradient within the pore (even though compartments 1 and 2 are at the same pressure) that drives water through the pore (Fig. 2-6A). If the system is brought to equilibrium by applying a hydrostatic pressure difference ($= RTc_s$) between compartments 1 and 2, then there is no pressure gradient within the pore, and hence no flow of water across the membrane, even though the hydrostatic pressure in compartment 1 is greater than that in compartment 2 (Fig. 2-6B).

The concept that osmotic flow across a porous membrane is driven by a pressure gradient within the pores was originally put forth by Vegard (1908) and subsequently rediscovered by Garby (1957) and Mauro (1957; 1960). A detailed kinetic analysis of the interfacial region, where the lower pressure within the pore develops, is probably not possible at this time, given present inadequacies in the theory of liquids; qualitatively, the drop in pressure across interface 1 in Figures 2-6A and 2-6B results from the failure of the impermeant solute to transfer its momentum to the pore contents (water) when it collides with the membrane (Mauro, 1957). Note that a lower pressure exists within the pore for the simple equilibrium situation in which compartments 1 and 2 contain equal concentrations of impermeant solute; for the chemical potential of water in the pore to equal that in the solutions, and thus for equilibrium to exist, the hydrostatic pressure within the pore must be lower than that in the solutions by RTc_s (Fig. 2-6C). Furthermore, as emphasized by

21

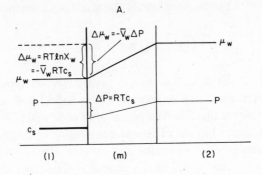

A.

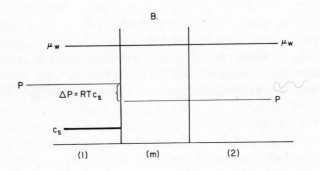

B.

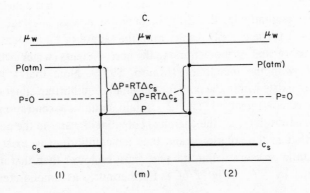

C.

22

Mauro (1965; 1981), this pressure will be negative, and a state of tension will exist within the pore, if the solute concentration is greater than 0.04 M.

An Aside on the Possible Effect of Osmotic Pressure on Channel Gating. Many biological channels undergo transitions between open and closed states. Their probability of being in a given state is often a function of some external parameter—such as membrane potential (e.g., voltage-dependent sodium channels in nerve) or a

---

Figure 2-6. Pressure and chemical potential profiles within the pores of a membrane separating a solution of impermeant solute (compartment 1) from pure solvent (compartment 2). (A) The pressure in both compartments is the same, and osmotic flow occurs. "Just within" the pores at interface 1, there is a pressure drop equal to RTc_s; this "discontinuity" in pressure compensates for the "discontinuity" in water mole fraction at that interface (caused by the "discontinuity" in solute concentration), and the chemical potential of water thereby remains continuous across the interface. As a consequence of the pressure drop at interface 1, a pressure gradient is set up within the pores, and it is this that drives water through the pores from compartment 2 to compartment 1. (B) The pressure in compartment 1 is greater than that in compartment 2 by RTc_s, and osmotic equilibrium exists. Note that there is still a pressure drop at interface 1 as in A, but this now lowers the pressure "just within" the pores at that interface to $P(2)$. Thus, there is no gradient of pressure (or of μ_w) within the pores, and hence no osmotic flow of water. (C) Illustration that there is a lower pressure within the pores for the simple equilibrium situation of the membrane separating equal concentrations of impermeant solute. At each interface there is a drop in pressure equal to RTc_s (written as RT Δc_s in the figure), and hence the pressure within the pores is lower than the ambient pressure by this amount. This lower pressure within the pores brings μ_w there to the value in the surrounding solutions, so that there is a constant value of μ_w throughout the system. In the example shown, c_s is greater than 0.04 M, and consequently the pressure within the pores is negative.

chemical ligand (e.g., cholinergic channels at the muscle endplate)—that is said to "gate" the channel. It has not been generally appreciated until recently that gating behavior can be affected by osmotic pressure (Zimmerberg and Parsegian, 1986). This will occur if there are volume changes associated with the transition of a channel from the open to closed state, in which case a PV term enters into the free energy of the transition.

To illustrate the possible magnitude of this effect, we envision a water-filled cylindrical channel of radius r and length L, and assume that in going from the open, conducting state (o) to the closed, nonconducting state (c), the channel lumen is obliterated throughout its length. In this transition a volume ($V = \pi r^2 L$) of solution is expelled from the channel into the surrounding medium, and conversely this volume enters the channel from the surrounding medium when the channel undergoes the transition from the closed to open state. There is thus a $V\Delta P$ term associated with the free energy change ($-\Delta G$) of the transition between the open and closed state of the channel. We can write for f, the ratio of time (t_o) spent in the open state to time (t_c) spent in the closed state (which, for an ensemble of channels is the ratio of the number of channels in the open state to the number in the closed state):

$$f \equiv \frac{t_o}{t_c} = \frac{[o]}{[c]} = \exp \frac{-\Delta G}{RT} = \exp \frac{-(\Delta G' - V\,\Delta P)}{RT}$$

$$= \exp \frac{-\Delta G'}{RT} \cdot \exp \frac{V\,\Delta P}{RT} \qquad (2\text{-}19)$$

where $-\Delta G'$ does not include the $V\Delta P$ contribution to the free energy in going from the open to the closed state, and ΔP is the hydrostatic pressure in the channel minus that in the surrounding medium.

Let us now calculate the contribution to f of the pressure term, $\exp(V\,\Delta P/RT)$, for a channel having the biological dimensions $r = 5$ $\overset{\circ}{A}$ and $L = 30$ $\overset{\circ}{A}$. [This corresponds to the estimated dimensions of that portion of the acetylcholine receptor channel that lies within the bilayer of the plasma membrane (Karlin, 1983).] If the surrounding medium does not contain an impermeant solute, $\Delta P = 0$ (i.e., the pressure in the channel is the same as that of the surrounding medium), and $V \Delta P = 0$ or $\exp(V \Delta P/RT) = 1$. That is, the pressure term makes no contribution to f. On the other hand, if the surrounding medium contains an impermeant solute at a concentration of 1 osmolar, $\Delta P = -24.6$ atm and therefore:

$$V \Delta P = -3.4 \times 10^{10} \text{ ergs/mole} \approx -1.36 \, RT$$

$$\therefore \exp \frac{V \Delta P}{RT} \approx 0.25$$

Thus, f is decreased about 4-fold when the osmolarity of the surrounding medium is raised to 1 molar with an impermeant solute; that is, the ratio of the time spent by the channel in the open state to that spent in the closed state is reduced by a factor of 4.

The effect of *osmotic pressure* on channel gating described above is very different from possible effects of *hydrostatic pressure* on gating. The former depends on volume changes in pore content in going from the open to closed state and in the *difference* in hydrostatic pressure between water within the pore and that in the surrounding medium (Fig. 2-7A). Externally applied hydrostatic pressure, on the other hand, is transmitted equally to water within and outside the pore (Fig. 2-7B), and it therefore does not alter the contribution made by volume changes in pore content to the free energy change of the transition between open and closed state. Any effects of externally applied hydrostatic pressure on the free energy of this transition will be associated with compressibility of, and small partial molar volume changes in, the protein (and lipid) forming the channel. Much larger hydrostatic pressures are required to produce measurable effects on channel gating than the osmotic pressures we have been considering (MacDonald, 1972; Suzuki and Taniguchi, 1972).

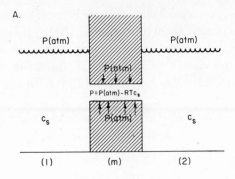

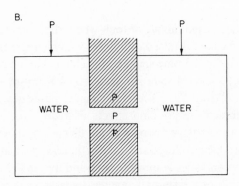

Figure 2-7. Illustration of how osmotic pressure could affect channel gating. The channel is shown in the open state; in the closed state, the channel volume (V) is assumed to be totally obliterated. (A) With equal concentrations of impermeant solute in both solutions, there is a lower pressure within the channel than elsewhere in the system. Note that there is an inward force on the walls of the channel that tends to drive it into the closed state. (B) There is no impermeant solute in the two solutions. The hydrostatic pressure has been raised in both solutions, but this is transmitted to the membrane matrix and the fluid within the channel. Consequently, there is no force acting on the walls of the channel, in contrast to the situation depicted in A.

OSMOSIS INDUCED BY AN IMPERMEANT SOLUTE

The Value of P_f

For an oil membrane, osmosis occurs by a solubility-diffusion mechanism, and P_f is directly related to the partition coefficient of water into the membrane phase and its diffusion constant within that phase (eq. 2-13). For a rigid porous membrane, we have seen that osmosis results from a hydrostatic pressure gradient ($\Delta P = RT\Delta c_s$) within the pore. Osmotic flow is thus indistinguishable from flow that would be generated by an applied hydrostatic pressure difference (ΔP) across the membrane in the absence of any Δc_s of impermeant solutes. If the pores are of macroscopic dimensions, this convective flow is characterized (for low Reynolds numbers) by a parabolic velocity profile within the pores, and its magnitude is given by Poiseuille's law:

$$J_v \equiv \Phi_w \overline{V}_w = n \frac{\pi r^4 \; \Delta P}{8L\eta} \tag{2-20}$$

where η is the viscosity of water. Comparing this with the definition of P_f in eq.(2-2) we obtain:

$$P_f = \frac{n}{A} \frac{\pi r^4 RT}{8L\eta \overline{V}_w} \quad \text{(for a membrane with macroscopic pores)} \tag{2-21}$$

There is no question that eq.(2-21) is correct for a membrane composed of large-radius pores, in which macroscopic hydrodynamic laws are applicable. Those laws pertain to a continuous fluid, that is, one in which differential elements of volume, though very small, contain a great many molecules. In other words, the finite size of solvent molecules can be neglected. It has been shown experimentally that eq.(2-20) [and hence eq.(2-21)] is still valid for pores of only 150 Å radius (Bean, 1972). The question of interest for physiologists and biophysicists, however, has been: what is the correct expression for P_f if the pore radii are of molecular dimensions, and therefore the finite size of water molecules cannot be neglected *a priori*? That is,

27

what equation replaces Poiseiulle's law [eq.(2-20)] for flow through narrow pores?

There are two modifications of eq.(2-20) most commonly introduced, particularly in calculating pore radii—a topic considered in Section B of Chapter 3. The first is the addition to Poiseuille's law of a "diffusive flux" term (see, for example, Pappenheimer, 1953; Durbin et al., 1956; Bean, 1972). The rationale for the inclusion of this term is the belief that, in addition to the laminar flow generated by a pressure difference, there is a diffusional component of flow resulting from the gradient in chemical potential of water (produced by the pressure gradient) within the pore. From the flux equation

$$\Phi_w = -u_w \, Ac_w \, \frac{d\mu_w}{dx} \tag{2-22}$$

where u_w is the mobility of water, we can write for this diffusional component of water flux through n pores:

$$\Phi_w = -u_w n \, \pi r^2 c_w \overline{V}_w \frac{dP}{dx} = -u_w(n\pi r^2)\frac{dP}{dx} = \frac{n\pi r^2 u_w \, \Delta P}{L}$$

and combining this with the Nernst-Einstein relation (Einstein, 1908):

$$D = uRT \tag{2-23}$$

it becomes

$$\Phi_w = \frac{n\pi r^2 D_w}{RT} \frac{\Delta P}{L}$$

where D_w is the diffusion coefficient of water within the pores. Adding this to eq.(2-20), we then have for the total magnitude of water flow (laminar plus diffusive) through the pores:

$$J_v = \frac{n\pi r^2}{L} \left(\frac{r^2}{8\eta} + \frac{D_w \bar{V}_w}{RT} \right) \Delta P \qquad (2\text{-}20a)$$

and P_f becomes:

$$P_f = \frac{n\pi r^2}{AL} \left(\frac{r^2 RT}{8\eta \bar{V}_w} + D_w \right) = \frac{n\pi r^2}{AL} \left(\frac{r^2 RT}{8\eta \bar{V}_w} \right) + P_{d_w} \qquad (2\text{-}21a)*$$

The second modification of eq.(2-20) [or (2-21)] is to alter the pore area (r^2) to an "effective" area for transport, so that one obtains (Paganelli and Solomon, 1957):

$$J_v = \frac{n\pi r^2}{L} \left[2\left(1 - \frac{a_w}{r}\right)^2 - \left(1 - \frac{a_w}{r}\right)^4 \right] \left[1 - 2.104 \left(\frac{a_w}{r}\right) + \right.$$
$$\left. 2.09 \left(\frac{a_w}{r}\right)^3 - 0.95 \left(\frac{a_w}{r}\right)^5 + \cdots \right] \left(\frac{r^2}{8\eta}\right) \Delta P$$
$$(2\text{-}20b)$$

$$P_f = \frac{n\pi r^2}{AL} \left[2\left(1 - \frac{a_w}{r}\right)^2 - \left(1 - \frac{a_w}{r}\right)^4 \right] \left[1 - 2.104 \left(\frac{a_w}{r}\right) + \right.$$
$$\left. 2.09 \left(\frac{a_w}{r}\right)^3 - 0.95 \left(\frac{a_w}{r}\right)^5 + \cdots \right] \left(\frac{r^2 RT}{8\eta \bar{V}_w}\right)$$
$$(2\text{-}21b)$$

where a_w is the radius of a water molecule. The first bracketed term is a correction for the reduced area available to water molecules, caused by steric restriction to their entry into the pore, and the second bracketed term corrects for the increased frictional interaction of finite-size (radius a_w) water molecules with the walls of the pore. Often the "diffusive flux" term, corrected for the "effective" area [see eq.(3-5a)], is added to eq.(2-21b) to give:

*See eq.(3-5) for P_{d_w}.

$$P_f = \frac{n\pi r^2}{AL}\left[2\left(1 - \frac{a_w}{r}\right)^2 - \left(1 - \frac{a_w}{r}\right)^4\right]\left[1 - 2.104\left(\frac{a_w}{r}\right) + \right.$$

$$\left. 2.09\left(\frac{a_w}{r}\right)^3 - 0.95\left(\frac{a_w}{r}\right)^5 + \cdots\right]\frac{r^2RT}{8\eta\overline{V}_w} + P_{d_w}$$

$$(2\text{-}21c)*$$

Both of the above-mentioned modifications of eqs.(2-20) and (2-21) lack good theoretical justifications. In the realm of large pores where the continuum assumptions are valid, all of the mass flow of water is accounted for in the derivation of Poiseuille's law, and it is therefore inappropriate to add an additional diffusive flux term, as in eq.(2-20a) (Levitt, 1973). In small pores of molecular dimensions, where the continuum assumptions are no longer valid, there must be a *de novo* derivation of the flow equations, and there is no theoretical reason to believe that its end result will be the simple patching up of Poiseuille's law with an additive diffusional term as in eq.(2-20a). The two bracketed terms in eqs.(2-20b) and (2-21b), the first from Ferry (1936) and the second from Faxén (see Renkin, 1954), which are intended as further palliatives for Poiseuille's law, come, themselves, from continuum theory and pertain to a dilute suspension of spheres in a fluid. [Actually, the second bracketed term, although applicable to *diffusion* of spheres, is not correct for convective flow, for which a different function of a_w/r applies (Bean, 1972).] Their applicability to the molecules of the fluid itself (i.e., water) in a narrow pore where the continuum assumptions are invalid has no theoretical justification. In sum, once one leaves the realm of large-radius pores where continuum theory (and, therefore, Poiseuille's law) applies, and enters the domain of pores with molecular dimensions,[†] there is no satisfactory theory to describe solvent flow generated by a hydrostatic or osmotic pressure gradient.

---

*See eq.(3-5a) for P_{d_w}.

[†] We exclude from discussion, for the time being, pores of such narrow dimensions that solvent molecules are in single-file array. The topic of single-file transport is considered separately in Chapter 4.

A very interesting approach to the study of convection through pores of molecular dimensions is that by Levitt (1973), who carries out molecular dynamics calculations for hard, smooth spheres of 1 Å radius moving through a pore of 3.2 Å radius. (It would have been more appropriate to model water molecules as spheres of 1.5 Å radius, instead of 1 Å radius; the corresponding radius of the pore would then be 4.8 Å, instead of 3.2 Å.) The remarkable result from these computer simulated "experiments" is that a pressure difference between the two ends of the pore gives rise to a flux of spheres (water molecules) that is almost identical to that predicted by Poiseuille's law alone! (The fluxes predicted by the commonly used modifications of Poiseuille's law for small pores, discussed above, are not in quite as good agreement with the "experimental" results.) Thus, an equation derived from continuum theory (Poiseuille's law) correctly predicts solvent fluxes through a model pore whose radius is only three times larger than that of the solvent molecules. We shall note again this phenomenon of the applicability of continuum theory to pores of molecular dimensions when we discuss transport through nystatin, amphotericin B, and gramicidin A pores in Part II.

3

Tracer Diffusion of Water and the Relationship Between P_f and P_{d_w}

In Chapter 2 we discussed membrane water permeability as measured by P_f, the osmotic or filtration permeability coefficient. This permeability coefficient, defined by eq.(2-2), is obtained from experiments in which bulk water flow is generated by an applied ΔP or Δc_s across the membrane. In this chapter we consider another measure of membrane water permeability. The membrane in this instance separates identical solutions; there is no ΔP or Δc_s across the membrane and hence no net flow. Instead, some of the H_2O in one compartment is replaced by isotopically labeled water (e.g., THO, DHO, or H_2O^{18}), and the transport rate across the membrane of isotopically labeled water (the tracer flux) is measured (Fig. 3-1). Of course, an equal flux of H_2O occurs in the opposite direction, and hence there is no net water flow across the membrane; there is simply tracer exchange.

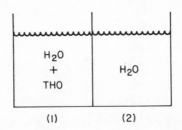

Figure 3-1. A membrane separates identical aqueous solutions, their only difference being that a small amount of the H_2O in solution 1 is replaced by THO.

We shall assume that there are no isotope effects; that is, the interaction of isotopically labeled water with the membrane and other water molecules is identical to that of H_2O. We also assume, as we tacitly did in our treatment of osmosis, that both solutions are well stirred, so that the concentration of molecules, in this case labeled water, at each membrane-solution interface is identical to that in the corresponding bulk solution. (In a subsequent section we discuss the problem of unstirred layers and their effects on osmotic and tracer flux measurements.) The flux Φ_w^* of isotopically labeled water across the membrane is then expressed by the equation:

$$\Phi_w^* = P_{d_w} A \Delta c_w^* \qquad (3\text{-}1)$$

where,

$$\Delta c_w^* \equiv c_w^*(1) - c_w^*(2) \qquad (3\text{-}2)$$

and P_{d_w} is the diffusion permeability coefficient of the membrane for water. P_{d_w} is another measure of a membrane's water permeability and is defined by eq.(3-1). The issue we now wish to address is: what is the relationship between P_f and P_{d_w}? The answer, as we shall see, is dependent on the nature of the membrane.

A. "OIL" MEMBRANE

By exactly the same arguments used in Section A of Chapter 2, we can write analogous to eq.(2-10):

$$\Phi_w^* = \frac{D_w A}{\delta} \Delta c_w^*(\text{oil phase})$$

and combining this with eqs.(2-6) and (2-7), where c_w and X_w are respectively replaced by c_w^* and X_w^*, we have:

$$\Phi_w^* = \frac{D_w K_w A}{\delta \overline{V}_{\text{oil}}} \Delta X_w^*(\text{water phase})$$

But since,

$$X_w^*(\text{water phase}) = c_w^*(\text{water phase})\overline{V}_w$$

we obtain,

$$\Phi_w^* = \frac{D_w K_w \overline{V}_w}{\delta \overline{V}_{\text{oil}}} A\Delta c_w^*$$

and comparing this with eq.(3-1) we finally have:

$$P_{d_w} = \frac{D_w K_w \overline{V}_w}{\delta \overline{V}_{\text{oil}}} \qquad \text{(for an oil membrane)} \qquad (3\text{-}3)$$

We thus see from eqs.(2-13) and (3-3) that:

$$\frac{P_f}{P_{d_w}} = 1 \qquad \text{(for an oil membrane)} \qquad (3\text{-}4)$$

The equality of P_f and P_{d_w} for an oil membrane is intuitively obvious. For both osmotically induced and tracer-water flux, water crosses the membrane by a solubility-diffusion mechanism. If the "oil" phase were water-rich, this would not be true for osmotically driven water flux; water molecules in the membrane would "see" each other and undergo a cooperative movement (as occurs, in the extreme, in a water-filled pore), and consequently $P_f \neq P_{d_w}$. The "oil" phase of lipid bilayers and plasma membranes, however, is much like hydrocarbon (see Chapter 6), in which water is poorly soluble, and it is for this reason that I have confined the discussion of oil membranes to those in which the concentration of water within the membrane phase is so low that interactions of H_2O molecules with each other can be neglected.

B. POROUS MEMBRANE (PORE RADIUS)

We begin our treatment of tracer-water flux through a porous membrane by first considering the case in which the pores are of macroscopic dimensions, and hence continuum theory is applicable. We can then obviously write, from Fick's law, for a membrane containing n right circular cylindrical pores of radius r and length L:

$$\Phi_w^* = \frac{n\pi r^2 D_w}{L} \Delta c_w^*$$

and by comparison with eq.(3-1) we obtain:

$$P_{d_w} = \frac{n}{A} \frac{\pi r^2 D_w}{L} \qquad \text{(for a membrane with macroscopic pores) (3-5)}$$

Dividing eq.(2-21) by eq.(3-5) we obtain:

$$\frac{P_f}{P_{d_w}} = \frac{RT}{8\eta D_w \bar{V}_w} r^2 \qquad (3\text{-}6)$$

or, if we use eq.(2-21a) rather than eq.(2-21), we have instead:

$$\frac{P_f}{P_{d_w}} = \frac{RT}{8\eta D_w \bar{V}_w} r^2 + 1 \qquad (3\text{-}6a)$$

At room temperature ($T = 298°$):

$$\frac{RT}{8\eta D_w \bar{V}_w} = 8.04 \times 10^{14} \text{ cm}^{-2} \qquad (3\text{-}7)$$

where we have used the values $D_w = 2.4 \times 10^{-5}$ cm²/s, $\eta = 0.89\,cP$, $\bar{V}_w = 18$ cm³/mole, and $R = 8.3 \times 10^7$ erg/deg/mole.

The proportionality of P_f/P_{d_w} to r^2 results from laminar flow being proportional to r^4, whereas diffusional flux is proportional to r^2; the larger the radius of the pores, the greater is the value of P_f/P_{d_w}. By combining osmotic flow (or pressure-induced flow) data with tracer flux data, one can calculate through eq.(3-6) the pore radius for a membrane composed of uniform right circular cylindrical pores. There are, of course, two caveats to the use of eq.(3-6). First, the pores in the membrane of interest may not be identical, and furthermore they may not be right circular cylinders. Ordinary dialysis tubing, for example, is assuredly porous in nature, yet it is not composed of uniform right circular cylindrical pores; in fact, a more accurate description of it would be a spaghetti-like meshwork. Despite this, however, values of P_f and P_{d_w} determined on dialysis tubing can be inserted into eq.(3-6) to calculate an "equivalent" pore radius for the membrane (Durbin, 1960). Although the number obtained cannot be interpreted literally, it does have heuristic value in predicting which solutes can and cannot cross the membrane.

The second caveat to the use of eq.(3-6) is that it is derived from continuum theory, whereas pores in biological membranes are generally of molecular dimensions. In Chapter 2, we discussed attempts to modify eq.(2-21), the expression for P_f obtained from continuum theory, to make it applicable to narrow pores. Here we consider similar attempts to modify eqs.(3-5) and (3-6) for applicability to P_{d_w} and P_f/P_{d_w} determinations for narrow pores. The term "narrow" probably pertains only to pores with radii a few times larger than that of a water molecule. It is noteworthy that equivalents of eqs.(2-21) and (3-5), and therefore also eq.(3-6), have been experimentally verified for membranes with 150 Å radius pores (Bean, 1972). Thus, continuum theory remains applicable to pores with radii only 100 times larger than that of a water molecule.

The commonly employed modification of eq.(3-5) is to regard the area of the pore (πr^2) as an "effective" area for diffusion and write (Renkin, 1954; Paganelli and Solomon, 1957):

$$P_{d_w} = \frac{nD_w}{AL}\pi r^2 \left(1 - \frac{a_w}{r}\right)^2 \left[1 - 2.104\left(\frac{a_w}{r}\right) + \right.$$

$$\left. 2.07\left(\frac{a_w}{r}\right)^3 - 0.95\left(\frac{a_w}{r}\right)^5 + \cdots \right] \quad (3\text{-}5a)$$

This modification of P_{d_w} is very similar to that used to modify P_f [see eq.(2-21b)] and suffers from the same lack of theoretical justification when applied to water molecules in a narrow pore, as discussed earlier for eq.(2-21b). Dividing eq.(2-21b) by eq.(3-5a) yields:

$$\frac{P_f}{P_{d_w}} = \left[2 - \left(1 - \frac{a_w}{r}\right)^2\right]\frac{RT}{8\eta D_w \overline{V}_w} r^2 \quad (3\text{-}6b)$$

More commonly, the presumed diffusional contribution to P_f, $n\pi r^2 D_w/AL \; (= P_{d_w})$, is included in P_f [see eq.(2-21a)], and, in fact, this contribution to P_f is written in the form of eq.(3-5a) [see eq.(2- 21c)]. Dividing eq.(2-21c) by eq.(3-5a) then gives:

$$\frac{P_f}{P_{d_w}} = \left[2 - \left(1 - \frac{a_w}{r}\right)^2\right]\left(\frac{RT}{8\eta D_w \overline{V}_w} r^2\right) + 1 \quad (3\text{-}6c)$$

Equations (3-6a) and (3-6c) have been extensively used to calculate equivalent pore radii for biological membranes (see Solomon, 1968), and for this reason they were presented here. It is generally acknowledged, however, as was emphasized in the discussion of their derivations, that they lack a theoretical foundation. In a way, these equations have taken on a life of their own, and equivalent pore radii are *defined* by the values coughed up when experimental determinations of P_f/P_{d_w} are shoved into them. There is presently no satisfactory physical theory for diffusion and flow through pores with radii only a few times larger than that of the solvent molecule. Paradoxically, however, there is a much better understanding for diffusion and flow (and, particularly, for the magnitude of P_f/P_{d_w}) through pores with radii so small ($<2a_w$) that single-file transport occurs within them. This topic is considered in the next chapter, but before moving on to it, we briefly digress to discuss a significant problem that can arise in experimental determinations of P_f and P_{d_w}.

C. UNSTIRRED LAYERS

Although radii of narrow pores cannot be calculated with any degree of confidence from eqs.(3-6a) and (3-6c), the value of P_f/P_{d_w} is important in its own right for deciding if the primary route for water flow across a plasma membrane is through pores or through the bilayer proper. In particular, if P_f/P_{d_w} is significantly greater than 1, one can conclude that pores provide a major pathway for water transport. On the other hand, if $P_f/P_{d_w} = 1$, one can surmise that water crosses the membrane through the bilayer proper by a solubility-diffusion mechanism.* It turns out, however, that measurements of P_f/P_{d_w} are often complicated by the presence of unstirred layers, or other barriers, in series with the membrane. In general, these have a greater effect on P_{d_w} than on P_f, and consequently spuriously high values of P_f/P_{d_w} can be obtained experimentally, even if P_f/P_{d_w} actually equals 1 for the membrane of interest. Let us see how this arises.

Given a membrane separating two stirred solutions, there will always be, near each membrane-solution interface, a region of incomplete mixing (Schulman and Teorell, 1938). The easiest, and generally most satisfactory, way of incorporating this fact into permeability calculations is to idealize it by assuming perfect mixing up to a certain distance δ from the membrane surface (δ_1 and δ_2 at membrane-solution interfaces 1 and 2, respectively), and no mixing of contents in the region between this plane and the membrane surface (Fig. 3-2).

---

*It might appear, from eqs.(3-6a) and (3-6c), that $P_f/P_{d_w} \approx 1$ for very narrow pores ($r \approx a_w$). In general, however, this will not be the case (see Chapter 4).

Consider first the case of tracer diffusion (Fig. 3-2A). There are three regions in series (the membrane itself and the two unstirred layers), and in the steady state, Φ_w^* must be the same throughout all three. In the unstirred layers, transport occurs strictly by diffusion; there is no convective component to the tracer flux. It is easy to show that $(P_{d_w})_{\text{obs}}$, the experimentally observed diffusion permeability coefficient, is related to P_{d_w}, the true diffusion permeability coefficient of the membrane, through the equation:

$$\frac{1}{(P_{d_w})_{\text{obs}}} = \frac{1}{P_{d_w}} + \frac{1}{D_w/\delta_1} + \frac{1}{D_w/\delta_2}$$

or

$$(P_{d_w})_{\text{obs}} = \frac{1}{1 + P_{d_w}(\delta/D_w)} P_{d_w} \qquad (3\text{-}8)$$

where $\delta = \delta_1 + \delta_2$.

Consider now the case of osmotic flow, and for simplicity let the impermeant solute s be present only in solution 1 (Fig. 3-2B). The flow of water through the unstirred layer dilutes the solute concentration at the membrane surface, and hence the effective osmotic gradient across the membrane is less than that applied. In the steady state, the flux (Φ_s) of solute in the unstirred layer is zero, and we can therefore write:

$$\Phi_s = D_s A \frac{dc_s}{dx} + vAc_s = 0$$

where v is the velocity of flow and c_s is the solute concentration at any point x in the unstirred layer. Integrating between $x = -\delta_1$ (the beginning of the unstirred layer) and $x = 0$ (the membrane surface), we obtain:

$$c_o = c_{\text{bulk}}\, e^{-v\delta_1/D_s}$$

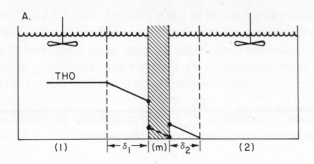

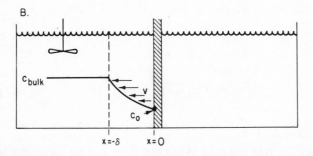

Figure 3-2. Effect of unstirred layers on tracer flux (A) and on osmotic flux (B). The thicknesses of the unstirred layers on the two-sides of the membrane are δ_1 and δ_2, respectively. (A) The concentration profiles of THO in the unstirred layers and within the membrane are shown. (For simplicity an oil membrane is shown in which it is assumed that the diffusion constant of water is the same as in aqueous solution.) In the steady state, the flux of water through each unstirred layer and through the membrane must be equal. Note that because of the unstirred layers, the actual concentration difference of THO across the membrane is less than that given from the values in the two solutions, and hence a lower flux of THO is observed (and a lower value of P_{d_w} is calculated) than would be observed in the absence of unstirred layers. Furthermore, the thinner the membrane and the thicker the unstirred layers, the greater is the discrepancy between $(P_{d_w})_{obs}$ and the true value of P_{d_w}. (B) The concentration profile of impermeant solute in the unstirred layer in the face of osmotic flow across the membrane. The solute is swept away from the

where c_{bulk} is the concentration of solute in the bulk of the solution and c_o is its concentration at the membrane surface. The osmotic flux Φ_w is in this case:

$$\Phi_w = -P_f A c_o = -(P_f e^{-v\delta_1/D_s}) A c_{bulk}$$

and we see that the experimentally observed osmotic permeability coefficient $(P_f)_{obs}$ is related to the true osmotic permeability co-efficient P_f through the equation:

$$(P_f)_{obs} = P_f e^{-v\delta_1/D_s} \qquad (3-9)$$

Note that for small osmotic flows, $(P_f)_{obs}$ approaches P_f. But small osmotic flows can always be achieved simply by applying a sufficiently small osmotic gradient. Thus, whereas the error in measuring P_{d_w} is fixed by the unstirred layer thickness [see eq.(3-8)], the error in measuring P_f can, in principle, be made as small as desired; in practice, it is usually trivial. For this reason, unstirred layers generally have a much larger effect on P_{d_w} than on P_f and, consequently, often lead to spuriously large values of P_f/P_{d_w} (Dainty, 1963).*

---

membrane interface by the fluid flow (of velocity v) through the unstirred layer, but the steady state, solute flux is zero, and its diffusional flux compensates for this solvent drag ($D_s\,dc_s/dx + vc_s = 0$) at all points in the unstirred layer. Note that the concentration of solute at the membrane interface (c_0) is lower than that in bulk solution (c_{bulk}), and hence the actual osmotic driving force (RTc_0) is less than that which would exist in the absence of an unstirred layer (RTc_{bulk}). Consequently, the observed osmotic flow (and observed P_f) is reduced.

*There is an additional factor that can cause unstirred layers to have a much smaller effect on P_f. The density gradient resulting from the concentration gradient of impermeant solute in the unstirred layer can lead to convective mixing there, and thus reduce the effective thickness of the "unstirred" layer (Everitt and Haydon, 1969).

4

Single-File Transport

This topic could have been addressed in Chapters 2 and 3, during our analyses of osmotic flow and tracer diffusion. I have chosen to deal with it in a separate chapter of its own, however, because the physics of single-file transport are unique. By single-file transport of water one means that water molecules within the pore cannot pass or overtake each other. [The term "no-pass" has been suggested as a better description of this mode of transport than "single-file" (Levitt, 1984).] This will certainly occur if the pore radius is less than twice that of a water molecule; conceivably, it could also occur in somewhat larger pores if forces other than strictly steric ones prevented water molecules from passing each other. Single-file transport of ions has received considerable attention in recent years (see Stevens and Tsien, 1979), but, for the most part, we shall confine our attention to water movement. In the last section of this chapter, however, we discuss ion-water coupling when both water and ions are restricted to the no-pass condition.

A. OSMOSIS (p_f)

Before discussing the nature of solvent flow resulting from an osmotic or hydrostatic pressure difference, we first derive an expression for p_f, the osmotic permeability coefficient per pore. p_f is related to the osmotic permeability coefficient for a membrane of area A containing n pores (P_f) by the expression:

$$p_f \equiv \frac{P_f A}{n} \qquad (4\text{-}1)$$

Figure 4-1. Pore, of length L, through which single-file transport of water molecules (shown as open circles) occurs.

(P_f has dimensions of cm/s; p_f has dimensions of cm³/s). The derivation is that suggested by Dr. Robert Myerson, formerly at the Institute for Advanced Study, and published by Finkelstein and Rosenberg (1979).

Consider a pore of length L, containing N water molecules, through which single-file transport occurs (Fig. 4-1). Let there be an osmotic pressure difference ($\Delta\Pi$) between the two solutions separated by the pore, given by

$$\Delta\Pi = kT\,\Delta n_s \qquad (4\text{-}2)$$

where k is the Boltzmann constant and Δn_s is the difference in impermeant solute concentration in the two solutions, expressed as molecules of solute per unit volume. [Equation (4-2) is identical to eq.(1-4b), since $k = R/N_A$ and $\Delta n_s \equiv N_A\Delta c_s$, where N_A is Avagadro's number.] The osmotic pressure difference $\Delta\Pi$ exerts a force (F_Π) on the contents of the pore completely analogous to that exerted by a hydrostatic pressure difference ΔP. The work W done by this force when N molecules cross the pore is simply PV work given by:

$$W = \bar{v}_w\,N\Delta\Pi$$

where $\bar{v}_w(= \bar{V}_w/N_A)$ is the volume per water molecule in bulk solution. But, from the definitions of work and force, we also have:

$$W = F_\Pi\,L$$

so that,

$$F_\Pi = \frac{\bar{v}_w \, N \, \Delta\Pi}{L} \qquad (4\text{-}3)$$

If the pore contents move with a velocity υ, the molecules experience a frictional drag F_γ, which can be written as:

$$F_\gamma = N\gamma\upsilon \qquad (4\text{-}4)$$

where γ is the frictional coefficient per water molecule. [Actually, γ is defined by eq.(4-4); it is simply the frictional force per unit velocity (F_γ/υ) divided by the number of water molecules in the pore (N).] For steady-state flow, which is what is always considered, F_Π is balanced by F_γ (i.e., $F_\Pi = -F_\gamma$), and we can therefore write from eqs.(4-2)–(4-4):

$$\upsilon = \frac{\bar{v}_w \, kT \, \Delta n_s}{\gamma L} \qquad (4\text{-}5)$$

The osmotic water flux Φ_w (in units of molecules per second) can then be written as:

$$\Phi_w = \frac{N\upsilon}{L} = -\frac{\bar{v}_w \, kTN}{\gamma L^2} \, \Delta n_s$$

and since, by definition, for a single pore:

$$\Phi_w \equiv -p_f \, \Delta n_s \qquad (4\text{-}6)$$

we finally arrive at an expression for p_f:

$$p_f = \frac{\bar{v}_w \, kTN}{\gamma L^2} \qquad (4\text{-}7)$$

A very similar expression is obtained from the work of Longuet-Higgins and Austin (1966), who carry out a statistical mechanical calculation for flow through a single-file pore. Their equation (4.9) can be cast in the form:

$$p_f = \frac{\bar{v}_w \, D_w^{(o)} \, N}{L^2} \tag{4-8}$$

where $D_w^{(o)}$ is the "diffusion coefficient" of a single water molecule in the pore and is obtained from the integral of the autocorrelation function of the velocity of the molecule. From the Nernst-Einstein relation [eq.(2-23)] we can write:

$$D_w^{(o)} = \frac{kT}{\gamma} \tag{4-9}*$$

and eq.(4-8) then becomes identical to eq.(4-7).

For osmotic or hydrostatic pressure-generated solvent transport through *"macroscopic" pores*, I emphasized (see Chapter 2) that solvent flow was *not* diffusive. In contrast, eq.(4-8) suggests that solvent flow *is* diffusive in nature for *narrow pores* in which single-file transport occurs. Indeed, if the density of water in the pore is the same as that in bulk solution, then:

$$N = \frac{\pi r^2 L}{\bar{v}_w}$$

(where r is the radius of the pore), and eq.(4-8) becomes:

$$p_f = \frac{\pi r^2 D_w^{(o)}}{L} \tag{4-8a}$$

---

*I shall comment later on this expression for $D_w^{(o)}$. It turns out that $D_w^{(o)}$, as defined by Longuet-Higgins and Austin (1966), is the diffusion coefficient a water molecule would have if it were the only molecule in the pore. It corresponds to D^o in the paper by Levitt and Subramanian (1974).

Equation (4-8a) is just the expression that would be written for p_f if one assumed that osmotic flow resulted from water molecules diffusing down their concentration gradient in the pore. In fact, it is precisely of the same form as P_{d_w} in eq.(3-5) for tracer diffusion through macroscopic pores.

Equations (3-6a, b, and c), which give expressions for P_f/P_{d_w} derived from continuum theory, indicate that the magnitude of this ratio declines monotonically as the radius of the pore decreases, and that it converges to a value close to 1 as the pore radius approaches that of a water molecule. Similarly, the discussion in the preceding paragraph of the diffusive nature of osmotic flow for single-file transport appears to indicate that $P_f/P_{d_w} = 1$ when the pore radius is less than twice that of a water molecule. Thus, the expectation from continuum theory that $P_f/P_{d_w} \approx 1$ as the pore radius becomes comparable to that of the solvent molecule seems to be independently confirmed by the analysis of single-file osmotic flow. In point of fact, however, there is not a convergence of continuum theory and single-file theory with respect to the value predicted for P_f/P_{d_w}. This we now demonstrate by obtaining an explicit expression for p_{d_w}. The derivation again follows Myerson's as published by Finkelstein and Rosenberg (1979).

B. TRACER DIFFUSION OF WATER (p_{d_w})

Consider again the same single-file pore of Figure 4-1, except in this instance let the pore contain a tracer water molecule; imagine also that an external force (F) is applied to that molecule alone (Fig. 4-2).

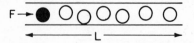

Figure 4-2. Pore, of length L, through which single-file transport of water molecules occurs. The pore contains one tracer water molecule (solid circle) and the external force (F) acts only on it.

(If the tracer is of a different density from that of ordinary water, the force could be gravitational.) In the steady state this force causes the entire pore contents to move with velocity v; since the applied force F is balanced by the frictional force F_γ, we have from eq.(4-4)

$$F = N\gamma v$$

If F were to act alone, it would give rise to a systematic flux ($\vec{\Phi}^*$) of tracer through the pore given by:

$$\vec{\Phi}^* = \frac{N^*v}{L} = \frac{N^*}{N}\frac{F}{\gamma L} \qquad (4\text{-}10)$$

where N^* is the mean number of tracers in the pore (and will be less than one for tracer quantities). On the other hand, if there were no applied force F, but instead a difference in tracer concentration Δn^* between the two solutions separated by the pore, there would be a diffusive flux Φ^* given by:

$$\Phi^* = p_{d_w} \Delta n^* \qquad (4\text{-}11)$$

where p_{d_w} is the diffusion permeability coefficient per pore for water and is related to P_{d_w} for a membrane of area A containing n pores through the equation:

$$p_{d_w} \equiv \frac{P_{d_w} A}{n} \qquad (4\text{-}12)$$

Now suppose that an equilibrium situation exists in which the applied force F and the phenomenological force resulting from the

47

Δn^* balance each other.[†] Then, following Einstein's (1905) treatment of diffusion in free solution, we view equilibrium as a balance of diffusive flux [eq.(4-11)] and systematic flux [eq.(4-10)]. That is:

$$\Phi^* = - \overrightarrow{\Phi}^*$$

and therefore,

$$- p_{d_w} \Delta n^* = \frac{N^*}{N} \frac{F}{\gamma L} \tag{4-13}$$

It only remains now for us to find, from equilibrium considerations, a relation between the applied force F and the thermal force (which is a function of Δn^*) that balances it. This expression will not involve p_{d_w} and γ (since it arises from equilibrium arguments and hence must be independent of transport coefficients); upon combining it with eq.(4-13) we shall obtain an explicit expression for p_{d_w} that is independent of Δn^* and F.

Calling the concentrations of tracer in the left and right compartments n_l^* and n_r^*, respectively, their ratio at equilibrium must satisfy the Boltzmann distribution:

$$\frac{n_l^*}{n_r^*} = e^{- FL/kT}$$

from which we obtain:

$$- \Delta n^* \equiv (n_r^* - n_l^*) = n_r^*(1 - e^{-FL/kT})$$

[†]To achieve thermodynamic equilibrium for both tracer and unlabeled water, a pressure difference equal to $kT \Delta n^*$ must also be established between the two ends of the pore. It is easy to show (see Levitt, 1974), however, that this pressure difference is small compared to $FL/N\bar{v}_w$ if, as in the derivation to follow, the applied force F, and hence Δn^*, is small, so that there is never more than one tracer molecule in the pore. In fact most of the time there will be no tracer in the pore.

For small forces, and hence small Δn^*, this expression becomes:

$$-\Delta n^* = \frac{FL}{kT} n^* \qquad (4\text{-}14)$$

where n^* is the average concentration of tracer in the two compartments and is defined by:

$$n^* \equiv \frac{n_l^* + n_r^*}{2}$$

At the value of Δn^* given by eq.(4-14), equilibrium exists, and the diffusive flux balances the systematic flux; that is, eq.(4-13) is satisfied. Combining eqs.(4-13) and (4-14) we obtain:

$$p_{d_w} = \frac{\bar{v}_w \, kT}{\gamma L^2} \frac{N^*}{N} \frac{1}{n^* \bar{v}_w} \qquad (4\text{-}15)$$

where we have multiplied numerator and denominator by $\bar{v}_w$. For a true tracer with no isotope effect, the fraction of molecules labeled in the bulk solution $(n^* \bar{v}_w)$ must equal the fraction labeled in the pore (N^*/N). Equation (4-15) thus simplifies to:

$$p_{d_w} = \frac{\bar{v}_w \, kT}{\gamma L^2} \qquad (4\text{-}16)$$

C. THE RATIO OF P_f TO P_{d_w}

If we divide eq.(4-7) by eq.(4-16), we obtain the remarkable result that:

$$\frac{P_f}{P_{d_w}} \equiv \frac{p_f}{p_{d_w}} = N \qquad \text{(for single-file transport)} \qquad (4\text{-}17)$$

For a narrow pore in which water molecules cannot pass each other, P_f/P_{d_w} equals the number of water molecules in the pore! This result is quite different from that anticipated from continuum theory [either eq.(3-6a, b, or c)] for pores with radii less than twice that of a water molecule (where the "no-pass" condition obtains). Figure 4-3 plots P_f/P_{d_w} as predicted by the most commonly used expression from continuum theory, eq.(3-6c), and compares it with the result from single-file theory [eq.(4-17)]. P_f/P_{d_w}, as described by continuum theory, is a monotonically decreasing function of pore radius that is about two for $r = 2a_w$. On the other hand, not only does single-file

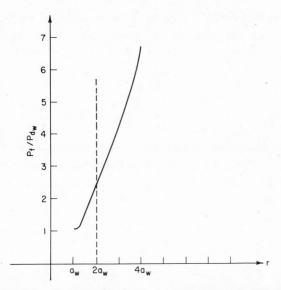

Figure 4-3. A sketch of P_f/P_{d_w} vs. pore radius (r) as given by eq. (3-6c). [The factor $RT/8\eta D_w \bar{V}_w$ in eq. (3-6c) is taken as 8.04×10^{14} cm^{-2} [see eq. (3-7)] and the radius of a water molecule (a_w) is assumed to be 1.5 Å.] Note that P_f/P_{d_w} monotonically decreases as r decreases, even in the region $a_w \leq r \leq 2a_w$, and is approximately equal to 1 when $r = a_w$. In contrast, single-file theory states [see eq. (4-17)] that $P_f/P_{d_w} = N$, the number of water molecules in the pore, when $r < 2a_w$.

theory not give the same value for P_f/P_{d_w} as predicted by continuum theory when $r < 2a_w$, it claims that this ratio is no longer even a *function* of pore radius; instead the value of P_f/P_{d_w} is a function of an entirely different parameter—the number of water molecules (N) in the pore, which all other things being equal, is proportional to pore length.

What accounts for this "discontinuity" in the behavior of the function P_f/P_{d_w}? Consider first eq.(4-7), the expression for p_f for single-file transport. For pores identical in all respects to the one in Figure 4-1 except for length, we can write, since $N \propto L$,

$$p_f \propto \frac{1}{L} \propto \frac{1}{N} \tag{4-18a}$$

The proportionality of p_f to $1/L$ (and hence to $1/N$) is in agreement with continuum theory; there is nothing remarkable, at least in this respect, about single-file solvent flow driven by a hydrostatic, or osmotic, pressure difference. In contrast, consider the implication of eq.(4-16), the expression for p_{d_w} for single-file transport. For pores identical in all respects to the one in Figure 4-1 except for length, we find, instead of relation (4-18a), that

$$p_{d_w} \propto \frac{1}{L^2} \propto \frac{1}{N^2} \tag{4-18b}$$

The proportionality of p_{d_w} to $1/L^2$ (and hence to $1/N^2$) is unique to single-file transport; in continuum theory, p_{d_w}, like p_f, is proportional to $1/L$. Thus, it is the unusual dependence of p_{d_w} on $1/L^2$ that causes P_f/P_{d_w} for single-file transport [eq.(4-17)] to differ so drastically from the predictions of continuum theory. If the denominator of eq.(4-16) contained L instead of L^2, P_f/P_{d_w} would equal 1 and be in reasonable agreement with the predictions of continuum theory.

Why does the dependence of p_{d_w} on pore length change from $1/L$, for a macroscopic pore, to $1/L^2$, for a single-file pore? Basically, it results from a fundamental change in the nature of the diffusion process. In a macroscopic pore (i.e., in a pore wide enough to accommodate several molecules side by side), a water molecule, or tracer water molecule, diffuses more or less as in free solution—hopping from one vacancy to the next, independent of similar hoppings by the other water molecules. Even in a double-file pore with one vacancy, a tracer molecule can travel from one end to the other without a required translocation of the other molecules in the pore (Fig. 4-4A). This is precisely what is precluded in single-file diffusion. In order for a tracer molecule to travel through the pore, all of the other molecules in the pore *must* be translocated along with it (Fig. 4-4B). This is the unique aspect of single-file diffusion. Notice that in the double-file pore, the vacancy can diffuse in step with the tracer (Fig. 4-4A), although of course, it can also wander off independently. In the single-file pore this cannot happen. After each step that the tracer takes, it *must* wait until the vacancy has diffused at

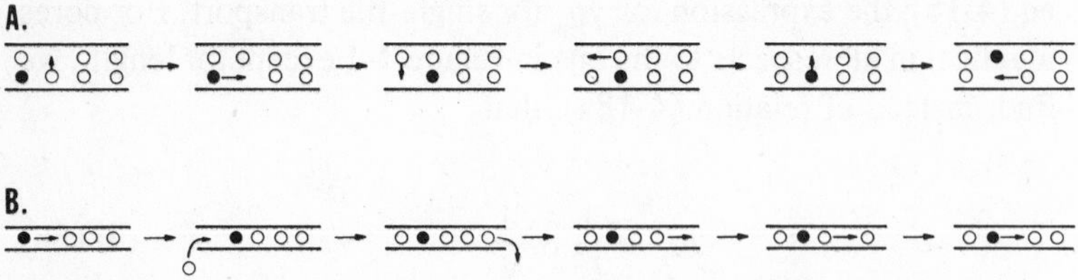

A.

B.

Figure 4-4. Diagram of the "hopping" of a tracer water molecule (solid circle) through a double-file pore (A) and a single-file pore (B) containing a single vacancy. Successive panels are sequential time frames. Note that in the double-file pore the tracer and the vacancy can move in step, whereas in the single-file pore, once the tracer has advanced one step, it must wait for the vacancy to move the entire length of the pore before it can advance another step. [In B, a water molecule is shown entering and leaving the pore (panels 2 and 3) as the vacancy reaches one end and reappears at the other end.]

least N steps, the entire length of the pore, before it can advance another step (Fig. 4-4B). Thus, in a sense, the rate at which the tracer advances one step is proportional to $1/N$, and therefore its rate of advancing N steps is proportional to $1/N^2$. This is the intuitive reason for the $1/N^2$-dependence of p_{d_w} in single-file diffusion.

The derivation of eq.(4-17), which is an exegesis of the one originally given by Levitt (1974), is very general. No assumptions are made about transport kinetics, other than that water molecules cannot pass each other within the pore (the "no-pass" or "single-file" condition). In particular, the derivation is independent of the physics of water molecule interactions with each other and with the pore wall. The derivation is based on an analysis of the balance of forces (and the fluxes produced by each force alone) in a system at thermodynamic equilibrium. It is basically an extension to a membrane system of Einstein's (1905) concept that the equilibrium distribution of particles in a conservative field (i.e., the Boltzmann distribution) can be thought of as being maintained by a balance of two competing fluxes.

There have been several other derivations of eq.(4-17), or variants thereof. Lea (1963) [see also Dick (1966)] analyzes a "knock-on" kinetic model of water movement through the pore; that is, the entry of a molecule from the adjacent solution at one end causes all N molecules in the pore to advance one step, thereby discharging the Nth molecule into the solution at the opposite end. He shows that a tracer molecule, entering the pore at one end, has a probability of $1/(N + 1)$ of traversing the pore and exiting at the opposite end. This calculation by Lea for tracer diffusion is straightforward, but it is unclear what mechanism for osmotic flow he has in mind that allows him to thereby conclude that $P_f/P_{d_w} = N + 1$, a result almost identical to eq.(4-17).

Dawson (1982) also arrives at the expression,

$$P_f/P_{d_w} = N + 1$$

from general considerations of tracer flux-ratio, combined with the assumption that the pore always contains N water molecules.* He analyzes a situation of zero net tracer flux, in which the tracer flux produced by an osmotic pressure difference is balanced by that produced by a tracer concentration gradient. Although this is not a true equilibrium for the entire system (unlike the situation we considered in which the flux produced by the tracer concentration gradient was balanced by the flux produced by a force acting *only* on the tracer[†]), Dawson feels he can still apply equilibrium arguments to those pores that at any instant contain a tracer. In reality, however, even for those pores there is not an equilibrium, despite net tracer flux being zero, and hence, his derivation is flawed.[‡]

Kohler and Heckmann (1979; 1980) treat the kinetics of unidirectional fluxes in single-file pores, through which the molecules move by hopping from occupied to vacant sites. They conclude that P_f/P_{d_w} can be as small as $N - 3$ and as large as N [the latter agreeing with eq.(4-17)]. Their result does not necessarily conflict with eq.(4-17), since the latter was derived assuming the pore always contains N water molecules, whereas the kinetic model of Kohler and Heckmann allows up to two vacancies (i.e., $N - 2$ water molecules in the pore). It is encouraging that an apparently detailed kinetic analysis of single-file transport yields a result in basic agreement with that derived from a kinetically-independent, balance of forces and fluxes argument— namely, the value of P_f/P_{d_w} is a linear function (with a slope of one) of the number of water molecules in the pore.

In contrast to the spirit of this statement, Manning (1975) concludes from his analysis of osmosis and tracer diffusion for single-

---

*Whether the right-hand side of eq.(4-17) should be N or $N + 1$ (or even $N - 1$) is a question of end effects. It involves the issue of whether a vacancy must be, or is, created at the end of the pore when a water molecule enters or leaves, and it depends upon the rates of these steps relative to the rate of translocation within the pore.

[†] The pressure gradient needed to bring this entire system to equilibrium introduced only a small contribution to the tracer flux and was, in fact, not even in the direction to oppose tracer flux generated by the concentration gradient (see footnote on page 48).

[‡] Levitt's (1984) attempt to rework Dawson's derivation is also flawed, because he neglects water flow that occurs through pores that are devoid of tracer at any instant.

file transport that $P_f/P_{d_w} = 1$. [He also states that Longuet-Higgins and Austin (1966) arrive at this same result, but the basis for this assertion is not clear, since those authors derive an expression only for p_f (see eq.(4-8)) and do not (at least explicitly) deal with the question of p_{d_w}.] The source of the discrepancy between Manning's (1975) finding that $p_f/p_{d_w} = 1$ and Levitt's (1974) result that $p_f/p_{d_w} = N$ is not obvious. Their lines of reasoning [the latter explicated in the derivation of eq.(4-17)] are very different, and both are quite subtle, making it difficult to resolve their points of departure. It appears that the discrepancy in their conclusions arises from Manning's assumption that the tracer diffusion coefficient is equal to kT/γ [see also Manning (1976)], whereas because of the coupling of tracer diffusion to the movement of the other $(N-1)$ water molecules in the pore, Levitt sets it equal to $kT/N\gamma$ [see also Levitt and Subramanian (1974)]. This is the crux of the issue, since basically if the tracer diffusion coefficient equals kT/γ, p_{d_w} is proportional to $1/N$ (just as p_f is), whereas if it equals $kT/N\gamma$, then p_{d_w} is proportional to $1/N^2$ [as obtained in eq.(4-16)]. The conclusions from Kohler and Heckmann's (1979; 1980) kinetic analysis, and the experimental results on the gramicidin A channel discussed in Chapter 8, support the position that the value of p_f/p_{d_w} for single-file transport is dependent on the number of water molecules in the pore and is *not* generally equal to 1.

D. THE COUPLING OF ION AND WATER FLUXES

Throughout this monograph, we have confined our discussion of osmosis to non-ionic solutions, because the introduction of electrolytes opens a Pandora's box of complexities. Among these complexities, few are more intractable than electrokinetic phenomena. It is therefore especially ironic that for one of the most difficult topics that we consider—single-file transport—electroosmosis is particularly easy to understand, and electroosmotic and streaming potential data have a very simple interpretation. For these reasons, and because the subject arises later in conjunction with experiments on the gramicidin

A channel, we close this chapter with a discussion of electroosmosis through a permselective pore in which all species within the pore, both ions and water, are subject to the no-pass condition; in particular, water molecules and ions cannot pass each other.

Consider a membrane composed of pores, ideally selective for univalent cations, in which single-file transport of both ions and water occurs. The membrane separates symmetric salt solutions, and current is passed (from an external pair of electrodes) through the pores. Furthermore, let the salt concentration be sufficiently small that a pore never contains more than one cation (Fig. 4-5). For a given number of moles of charge (coulombs = current $\times$ time) crossing the membrane, a certain number of moles of water (volume) also cross. This is electroosmosis. Because of the constraint imposed by single-file transport, all N water molecules within a pore must accompany each ion that moves through it, and hence, N can be directly obtained from an electroosmotic experiment by dividing the number of molecules of water that flow through the membrane by the number of ions that produced that flow. That is, if J_w is the electro-osmotic water flow (in cc/s) produced by a current I, then

$$N = \frac{F}{I} \frac{J_w}{\overline{V}_w} \qquad (4\text{-}19)^*$$

where F is the Faraday.

Thus, if a membrane contains any number of identical perm-selective, single-file pores, the number of water molecules in *one* such pore can be determined from *macroscopic* data by two completely independent means. One is by eq.(4-17), from P_f and P_{d_w} values measured in separate osmotic and tracer flux experiments, respectively; the other is by eq.(4-19), from the results of an electroosmotic experiment. Note that the electroosmotic experiment gives the

---

*In practice, one usually measures J_v, the total volume flow, which includes, in addition to J_w, the volume associated with the ions themselves. This must be taken into account in calculating N from experimental data (see Levitt et al., 1978).

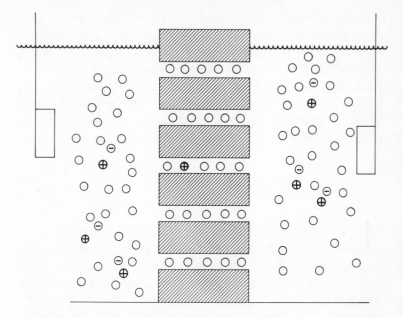

Figure 4-5. A membrane composed of pores that are ideally selective for univalent cations (⊕) and through which single-file transport of both ions and water molecules (o) occurs. The symmetric salt solutions separated by the membrane are sufficiently dilute that most pores do not contain a cation, and those that do, like the one shown here, have only one in them. A pair of electrodes in the two solutions is used for passing current through the membrane (the ion-containing pores) and thereby producing electroosmosis.

number of water molecules in an ion-containing pore, whereas, if the salt concentration is so small that at any instant most pores do not contain an ion, the P_f/P_{d_w} measurement gives the number of water molecules in an ion-empty pore. This point is elaborated upon in the discussion of the gramicidin A channel in Chapter 8.

We have focused on electroosmosis as the example of an electrokinetic measurement, because it is especially easy to see how N relates to it. It is also possible, however, to determine N from a streaming potential measurement. This can be demonstrated either from a direct consideration of streaming potential experiments, or from the phenomenological equivalence (as proved from irreversible thermodynamics) of electroosmotic and streaming potential data. By either approach, it can be shown (see Rosenberg and Finkelstein, 1978a; Levitt et al., 1978; Finkelstein and Rosenberg, 1979) that the streaming potential, $\Psi_{\text{streaming}}$, produced by an osmotic pressure difference $\Delta\Pi$ across the membrane is related to N by:

$$N = \frac{\Psi_{\text{streaming}}}{\Delta\Pi} \frac{F}{\overline{V}_w} \qquad (4\text{-}20)$$

Thus, the number of water molecules in an ion-containing, permselective pore obeying single-file kinetics can be determined from an electrokinetic measurement, which can come either from an electroosmotic experiment [eq.(4-19)] or from a streaming potential experiment [eq.(4-20)].

5

Osmotic Transport (Osmosis)
Induced by a Permeant (Leaky) Solute

Until now we have confined our discussion of osmosis to flow induced by a *totally impermeant* solute. In this chapter, we turn our attention to osmotic phenomena associated with a permeant or "leaky" solute. Actually, the diffusion of isotopically labeled water across membranes, the topic of Chapter 3, is a special case of this problem; the labeled water is a permeant "solute" identical to solvent molecules in all physical properties relevant to membrane permeation. Here we treat the general problem of the leaky solute, which has as its two limiting cases isotopically labeled water and totally impermeant solute.

The topic of combined, and interacting, solvent and solute fluxes falls naturally under the aegis of irreversible thermodynamics and is normally presented in the context of that formalism. That shall not be the approach taken here. Although irreversible thermodynamics has made some positive contributions to the subject of osmotic flow across membranes, particularly in lending its imprimatur to certain general results and conclusions, it has also, in my opinion, been a source for some serious confusions, and has acted as a fig leaf behind which underlying physical mechanisms are hidden. Cross (or coupling) coefficients are not a substitute for physical insight, as we shall discover at the end of this chapter, when, for the sake of completeness, the methods and results of irreversible thermodynamics are reviewed with respect to the leaky solute problem. There, a very misleading picture of the physics of transport emerges—one in which even elementary cause and effect relationships are inverted—when reliance

59

is placed in these coupling coefficients for enlightenment on the *actual* (as opposed to the *formal*) driving force for solvent flow. In keeping with the spirit of the previous chapters, we shall not shy from crawling around inside the membrane to view the concentration and pressure profiles affecting solute and solvent transport.

The situation under discussion is the same as that depicted in Figure 1-1, a membrane separating two aqueous solutions, one of pure water, the other of a solute at concentration c_s—except in this instance the membrane is permeable to both the solvent (water) and the solute. (More generally, the solute is present on both sides of the membrane, and its concentration difference is Δc_s). The equilibrium state for this system is trivially obvious: water and solute flow across the membrane until the solute concentration is the same in both compartments, at which point net flux of solvent and solute ceases and the system is at thermodynamic equilibrium. It is not possible, as for the totally impermeant solute, to achieve equilibrium by establishing a hydrostatic pressure difference across the membrane, because one cannot equate by this means the chemical potentials of both water *and* solute in the two compartments. Nevertheless, one can ask the following question. Suppose both compartments are large and well stirred, so that during the course of an experiment the solute concentrations in the two compartments remain essentially constant, what hydrostatic pressure difference (ΔP) must be applied across the membrane to stop volume flow? Or, alternatively, what is the volume flow rate (J_v) across the membrane when there is no hydrostatic pressure difference applied ($\Delta P = 0$), and how does this rate compare to that which occurs if the solute is impermeant?

We shall see that the answer to the first question is that

$$\Delta P = \sigma \Delta \Pi = \sigma RT \, \Delta c_s$$

(pressure difference that must be applied to stop volume flow) (5-1a)

where σ is called the "reflection coefficient," and that the answer to the second question is

$$J_v = -\sigma L_p \, \Delta\Pi = -\sigma L_p RT \, \Delta c_s$$
(volume flow in the absence of ΔP) (5-1b)

If both a ΔP and a Δc_s are present,

$$J_v = L_p \, (\Delta P - \sigma \, \Delta\Pi) = L_p \, (\Delta P - \sigma RT \, \Delta c_s) \qquad (5\text{-}1)$$

instead of eq.(2-1b) for an impermeant solute. [Note that eqs.(5-1a) and (5-1b) are special cases of eq.(5-1) for $J_v = 0$ and $\Delta P = 0$, respectively.] The reflection coefficient σ can take on any value less than or equal to 1. For a totally impermeant solute, $\sigma = 1$; eq.(5-1) then reduces to eq.(2-1b) and eq.(5-1a) reduces to the equilibrium, van't Hoff expression [eq.(1-4b)]. For a permeant solute, σ is less than 1 and can even be negative, if the membrane is more permeable to the solute than to the solvent.

The reflection coefficient σ was introduced by Staverman (1951) from irreversible thermodynamic considerations discussed in the last section of this chapter. Our present objective is to gain some physical insight into its meaning. Why, if the solute is permeant, is a pressure less than $\Delta\Pi$ required to stop volume flow [eq.(5-1a)], and why is the flow rate induced by a $\Delta\Pi$ of permeant solute [eq.(5-1b)] less than that induced by a $\Delta\Pi$ of an impermeant one? The answers to these questions, as to any transport questions, are dependent on the membrane composition. As in our treatment of volume flow induced by an impermeant solute, we again focus on the same two classes of membranes: (1) lipoidal or "oil" membranes and (2) porous membranes.

A. "OIL" MEMBRANE

Volume flow J_v across an oil membrane is particularly simple. We make the same assumption for both water and solute (s) that we made previously for water alone, when it was the only permeant species. Namely, interfacial kinetics are rapid compared to diffusion through the membrane phase, so that essentially equilibrium conditions exist at the boundaries. If the solutions are chosen to be sufficiently dilute that the solute concentration in the membrane is small,* then water and solute diffuse independently in the membrane phase [they do not see each other (Fig. 5-1)], and we can therefore write for their respective fluxes in the absence of an applied ΔP,

$$\Phi_w = \frac{D_w A}{\delta} \Delta c_w (\text{oil phase}) \qquad (2\text{-}10)$$

$$\Phi_s = \frac{D_s A}{\delta} \Delta c_s (\text{oil phase}) \qquad (5\text{-}2)$$

In terms of concentrations in the aqueous solutions, these equations become:

$$\Phi_w = -P_f A \, \Delta c_s \qquad (2\text{-}12)$$

$$\Phi_s = P_{d_s} A \, \Delta c_s \qquad (5\text{-}3)$$

where,

$$P_f = \frac{D_w K_w \overline{V}_w}{\delta \overline{V}_{\text{oil}}} \qquad (2\text{-}13)$$

$$P_{d_s} = \frac{D_s K_s}{\delta} \qquad (5\text{-}4)$$

$$K_s \equiv \frac{c_s(\text{oil phase})}{c_s(\text{water phase})} \qquad (5\text{-}5)$$

---

*If the solute is lipophilic, its concentration in the membrane can be many fold larger than that in the solutions. Nevertheless, the latter can be chosen sufficiently dilute that the solute concentration in the membrane is small.

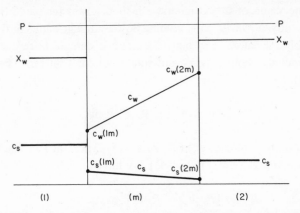

Figure 5-1. Concentration profiles of water (c_w) and solute (c_s) within an oil membrane separating solutions of unequal concentration of a permeant solute. (The solute is assumed to be more hydrophilic than water, so its partition coefficient, K_s, is much less than 1, and also less than that of water.) Water and solute diffuse independently through the membrane down their respective concentration gradients—water moving from compartment 2 to compartment 1 and solute moving in the opposite direction. Since this membrane is assumed to be considerably more permeable to water than to solute (the concentration gradient of water in the membrane is much steeper than that of the solute), net volume flow (osmosis) is from compartment 2 to compartment 1. The net volume flow from compartment 2 to compartment 1 is less than if the solute were impermeant (that is, if it did not partition into the membrane) by the amount of volume flow associated with the diffusion of solute from compartment 1 to compartment 2.

Water and solute cross the membrane independently by a solubility-diffusion mechanism, and the derivation for the preceeding equations follows the same lines as those in Chapter 2 when only water partitioned into the membrane phase.

If osmosis were defined in terms of solvent flow [that is, $J_w \ (= \bar{V}_w \Phi_w)$], then the osmotic flow through an oil membrane induced by a leaky solute would be equal to that induced by an impermeant solute, since the value of Φ_w is identical in both cases [eq.(2-12)]. Experimentally, however, total volume flow J_v, not just solvent flow, is invariably measured, and hence osmosis is defined in terms of *it*, which includes both solvent and solute volume transfer. Thus, by definition:

$$J_v \equiv J_w + J_s = \bar{V}_w \, \Phi_w + \bar{V}_s \, \Phi_s \qquad (5\text{-}6)$$

and substituting into this eqs.(2-12) and (5-3) plus the fact that for an oil membrane $P_f = P_{d_w}$ [eq.(3-4)], we have

$$J_v = -[P_{d_w}\bar{V}_w - P_{d_s}\bar{V}_s] \, A \, \Delta c_s \qquad (5\text{-}7)$$

Osmotic flow across an oil membrane is proportional to the difference in diffusion permeability coefficients (P_d's) for water and solute (corrected by their partial molar volumes). If the solute is impermeant, $P_{d_s} = 0$ and eq.(5-7) reduces to eq.(2-2) (less the ΔP term, set equal to zero in this example). On the other hand, if the membrane is more permeable to solute than to water (as would occur for a lipophilic solute), the sign of J_v is reversed. That is, the rate of volume transfer by solvent is less than the opposite rate of volume transfer by solute; there is "negative osmosis."

Using eqs.(2-3) and (3-4), we can rewrite eq.(5-7) as:*

$$J_v = -\left[L_p RT \, \Delta c_s - \frac{P_{d_s}}{RT} \bar{V}_s ART \, \Delta c_s \right]$$

and defining:

$$\omega \equiv \frac{P_{d_s}A}{RT} \qquad (5\text{-}8)$$

*Actually L_p is given by: $P_f A \bar{V}_w/RT + P_{d_s} A \bar{V}_s \bar{c}_s \bar{V}_s/RT$, instead of just by $P_f A \bar{V}_w/RT$ [see eq.(5-12)], where $\bar{c}_s \equiv (c_1 + c_2)/2$. But since we are confining ourselves to small $\bar{c}_s$, L_p reduces to $P_f A \bar{V}_w/RT$.

this becomes:

$$J_v = -L_p\left(1 - \frac{\omega \overline{V}_s}{L_p}\right) RT \, \Delta c_s \qquad \text{(for } \Delta P = 0) \qquad (5\text{-}9)$$

Comparing this with eq.(5-1b) we have:

$$\sigma = \left(1 - \frac{\omega \overline{V}_s}{L_p}\right) = \left(1 - \frac{P_{d_s}\overline{V}_s}{P_{d_w}\overline{V}_w}\right) \qquad (5\text{-}10)$$

(for an "oil" membrane)

Osmosis induced by a permeant solute is particularly simple to understand for an "oil" membrane. Essentially, permeant and impermeant solute induce the same *solvent* flow, but because there is volume flux in the opposite direction associated with diffusion of the permeant solute across the membrane, the *total volume flow* J_v induced by it is less. The expression for σ [eq.(5-10)] makes this point directly; σ differs from 1, its value for an impermeant solute, only by the term $\omega \overline{V}_s/L_p$. This is just $P_{d_s}\overline{V}_s/P_{d_w}\overline{V}_w$, the ratio of the "volume permeability coefficient" of the solute to that of water. For a polar solute that is much less permeant than water through an oil membrane, $\omega \overline{V}_s/L_p \ll 1$, and $\sigma \approx 1$. For isotopically labeled water as the "solute," $\omega \overline{V}_s/L_p = 1$, and $\sigma = 0$. For a lipophilic solute that is much more permeant than water through an oil membrane, $\omega \overline{V}_s/L_p > 1$, and $\sigma < 0$.

By exactly the same arguments presented in Chapter 2 for an impermeant solute, it is easily shown that a ΔP applied across an oil membrane separating symmetric solutions of a permeant solute gives rise to a diffusive flux of both water and solute, so that we can write for the flux of each:

$$\Phi_w = P_f A \, \frac{\Delta P}{RT} \qquad (2\text{-}17)$$

$$\Phi_s = P_{d_s} A \; c_s \overline{V}_s \frac{\Delta P}{RT} \tag{5-11}$$

Substituting these equations into eq.(5-6) along with eq.(2-3), we have:

$$J_v = L_p \left(1 + \frac{P_{d_s} \overline{V}_s A c_s \overline{V}_s}{L_p RT} \right) \Delta P \tag{5-12}$$

For the dilute solution case under consideration, c_s is negligible, and hence eq.(5-12) simplifies to:

$$J_v = L_p \, \Delta P \qquad \text{(for } \Delta c_s = 0) \tag{5-13}$$

A comparison of eq.(5-13) with eqs.(5-9) and (5-10) explicitly demonstrates for an "oil" membrane the validity of the general phenomenological statement in eq.(5-1): namely, the same coefficient L_p operates both on ΔP and on $\sigma \Delta \Pi$. That is, a ΔP gives rise to the same volume flow as a $\sigma RT \, \Delta c_s$. When ΔP equals $\sigma RT \, \Delta c_s$, volume flow is zero; at this pressure, the volume transported across the membrane by solvent from side 1 to side 2 equals the volume transported by solute in the opposite direction (Fig. 5-2).

B. POROUS MEMBRANE

Single-file Pore

Since it is relatively easy to get an intuitive sense for osmotic flow induced by a permeant solute through pores in which the no-pass condition applies, we discuss this case first, before turning in the next section to the more difficult problem of a membrane containing macroscopic pores.

Consider again the situation in Figure 1-1 in which the membrane separates a solution of permeant solute from one of pure solvent.

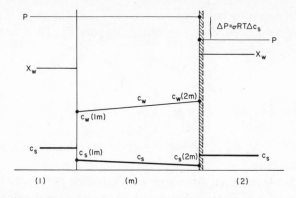

Figure 5-2. The concentration profiles of water and solute within an oil membrane when a ΔP equal to $\sigma RT \, \Delta c_s$ is applied across the membrane, thereby stopping volume flow. This figure should be compared to Figure 2-4, in which the solute is impermeant, and hence the ΔP ($= RT\Delta c_s$) establishes osmotic equilibrium. In that instance there is true equilibrium, no volume flow occurs, because there is no concentration gradient of water within the membrane. In the present case, there is not equilibrium; the zero volume condition results from equal and opposite rates of volume transfer by water and solute. (The concentration gradients of water and solute within the membrane are almost equal, having been brought to that condition by the ΔP; they differ only insofar as $D_w \bar{V}_w / D_s \bar{V}_s \neq 1$, where D_w and D_s are the diffusion coefficients in the membrane of water and solute, respectively.)

Furthermore, let the solution be so dilute that a pore contains at most one solute molecule; in fact, at any given time most pores have no solute in them. One can then roughly view volume movements at any instant as follows: through the solute-empty pores, water flows osmotically toward compartment 1 at the same rate as if the solute were impermeant. Through the solute-containing pores, solute diffuses toward compartment 2, and, because of the no-pass condition, carries with it the total volume of the pore contents, $\bar{v}_p^s$. This is analogous to the situation just considered for the "oil" membrane, with water and solute moving independently, except that instead of the solute transporting only its own volume $\bar{v}_s$, it transports the total volume (in solution) of a solute-containing pore, $\bar{v}_p^s$. Thus, instead of eq.(5-10), we have:

$$\sigma = \left(1 - \frac{\omega \bar{V}_p^s}{L_p} \right)$$

(5-14)

(if solute cannot pass solvent molecules)

where $\bar{V}_p^s \equiv N_A \, \bar{v}_p^s$ is the molar volume (in solution) of a solute-containing pore.

Eq.(5-14) has been given a proper derivation by Levitt (1974); here, we have contented ourselves with an intuitive development. Notice that in arriving at eq.(5-14) we invoked the no-pass condition only for solute with respect to solvent; the no-pass condition was not imposed on the solvent molecules themselves. Thus, eq.(5-14) also pertains to a membrane containing macroscopic pores, when the solute radius is so close to that of the pore radius that water molecules within the pore cannot slip around the solute. For pores so narrow that the no-pass condition also applies to the water molecules themselves, we have upon combining eqs.(4-17), (2-3), and (5-8) with eq.(5-14):

$$\sigma = \left(1 - \frac{P_{d_s}}{P_{d_w}} \frac{\bar{V}_p^s}{\bar{V}_p} \right) \qquad \text{(single-file pore)}$$

(5-15)

where $\bar{V}_p \equiv N\bar{V}_w$ is the molar volume (in solution) of a solute-free pore. Note that if the "solute" is isotopically labeled water, eq.(5-15) reduces to the expected result that $\sigma = 0$ (since $P_{d_s} = P_{d_w}$ and $\bar{V}_p^s = \bar{V}_p$).

Macroscopic Pore

One-dimensional Analysis. We begin by examining the concentration and pressure profiles within a pore when the membrane separates symmetric solutions of the permeant solute at concentration c_s(bulk) (Fig. 5-3). We considered the related situation for an

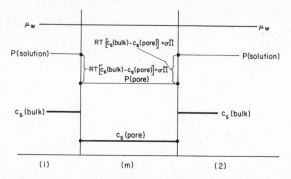

Figure 5-3. One-dimensional analysis of profiles within the pores of a membrane separating symmetric solutions of a permeant solute. This figure is completely analogous to Figure 2-6C, which should be consulted. The only difference arises because of the finite concentration, c_s(pore), of solute within the pores. As a consequence of this, the pressure drop within the pores, $RT[c_s(bulk)-c_s(pore)]$, is not as great as when the solute is impermeant, $RTc_s(bulk)$.

impermeant solute in Chapter 2 (Fig. 2-6C). There, for the chemical potential of water in the pore to equal that in the surrounding solutions, the pressure within the pore had to be lower than the ambient pressure; in fact,

$$P(\text{solutions}) - P(\text{pore}) = RTc_s(\text{bulk}) = \Pi$$
$$\text{(for impermeant solute)} \quad (5\text{-}16)$$

The present situation is analogous, except that instead of the solute concentration within the pore being zero, it is some finite value, $c_s(\text{pore}) < c_s(\text{bulk})$. Equating, as before, the chemical potential of water inside and outside the pore,* we obtain:

$$P(\text{solutions}) - P(\text{pore}) = RT[c_s(\text{bulk}) - c_s(\text{pore})] = \sigma\Pi \quad (5\text{-}17)$$

*The chemical potential of solute inside and outside the pore must also be equal. This point is discussed later.

CHAPTER 5

where,

$$\sigma = 1 - \frac{c_s(\text{pore})}{c_s(\text{bulk})} = 1 - K_s \qquad (5\text{-}18)$$

and,

$$K_s \equiv \frac{c_s(\text{pore})}{c_s(\text{bulk})} \qquad (5\text{-}19)$$

K_s is the partition coefficient of the solute between pore and bulk solution. The pressure within the pore is again lower than that of the surrounding solutions, but not as low as for an impermeant solute $[c_s(\text{pore}) = 0$, and therefore $\sigma = 1]$. In fact, for the limiting case of isotopically labeled water, $c_s(\text{pore}) = c_s(\text{bulk}) [\sigma = 0]$, and therefore the pressure within the pore is the same as that of the surrounding solutions.

If, now, the solute concentrations on the two sides of the membrane are *unequal,* it is clear from the preceding analysis that even though the bathing solutions are at the same pressure, there is a pressure difference between the ends of the pore. Under our usual assumption of equilibrium conditions at the boundaries, this pressure difference becomes:

$$(\Delta P)_{\text{in pore}} = RT\,\Delta c_s(\text{pore}) = \sigma RT\,\Delta c_s(\text{bulk}) = \sigma \Delta \Pi \quad (5\text{-}20)$$

and the resulting pressure gradient within the pore drives solution through the pore (membrane) from one compartment to the other (see Fig. 5-4A). This same result was obtained in Chapter 2 for the impermeant solute, except in that case the pressure difference between the ends of the pore was $\Delta \Pi$ ($\sigma = 1$). Osmotic flow is stopped (that is, $J_v = 0$) by an external ΔP equal to $\sigma \Delta \Pi$ (Figure 5-4B). In this stationary state, no net volume transfer occurs across the membrane; only interdiffusion of solute and water takes places.* The above considerations indicate that, for a proous membrane, flow produced by a ΔP is equivalent to that produced by a $\sigma RT \Delta c_s$.[†] Once again, therefore, the validity of the general phenomenological equation:

$$J_v = L_p (\Delta P - \sigma RT \Delta c_s) \qquad (5\text{-}1)$$

is confirmed for a particular type of membrane.

The derivation of eq.(5-20), and the analysis preceeding it, follow the approach taken by Garby (1957). The important conclusion drawn from that approach is that osmotic flow, induced by a Δc_s(bulk) of permeant solute across a membrane composed of macroscopic pores, is a convective, hydrodynamic flow indistinguishable from that produced by an externally applied hydrostatic pressure difference.[†] Indeed, flow generated by either a Δc_s(bulk) or an external ΔP is driven by a pressure gradient within the pore, and a volume element within the pore experiences the same force in both instances. The only difference, insofar as flow rate is concerned, between a permeant and impermeant solute is in the magnitude of the pressure gradient created within the pore by a given Δc_s(bulk).

Garby's approch, though elementary, properly focuses on the change in solute concentration at the solution-pore interface as the source of the pressure gradient within the pore. What is missing from that approach, however, is any explanation of why, given that it can enter the pore, the solute's concentration is lower there than in the adjacent bulk solution. Returning to the equilibrium situation of symmetric solutions (Fig. 5-3), how can the chemical potential of the *solute* within the pore and in the surrounding solutions be the same (as

---

*We neglect the small volume transfer that can result from a slightly unbalanced diffusive flow, due to unequal solute and solvent partial molar volumes (Hartley and Crank, 1949).

[†] The three-dimensional analysis, in the following section, of osmotic flow induced in macroscopic pores by a permeant solute leads to a modification of this conclusion.

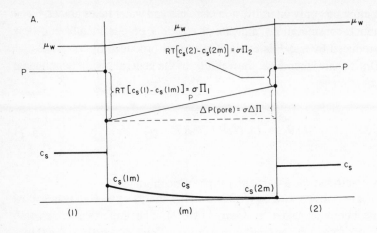

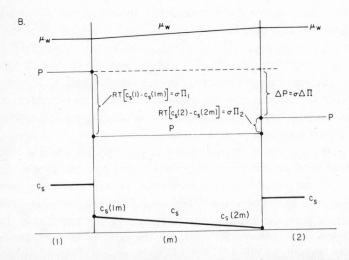

Figure 5-4. One-dimensional analysis of profiles within the pores of a membrane separating solutions of different concentrations of a permeant solute. (A) The pressure in both compartments is the same and osmotic flow occurs. This figure is analogous to Figure 2-6A, which should be consulted. (In that figure, compartment 2 contained pure water, instead of solute at a lower concentration than in compartment 1, but this is not an

it must be at equilibrium) if its concentrations in the two phases are different? To answer this question, and at the same time to gain a deeper insight into osmotic flow generated by a permeant solute, we turn to the more sophisticated analysis of Anderson and Malone (1974).

Three-dimensional Analysis. In contrast to Garby's one-dimensional view of solute concentration, Anderson and Malone (1974) emphasize the solute's three-dimensional distribution within the pore. From their general analysis of the problem, we confine our attention to their treatment of a membrane composed of uniform right circular cylindrical pores. In addition, the solute molecules are assumed to be hard spheres that interact only sterically with the pore walls. Thus, for a solute molecule of radius a_s, its potential energy of interaction, $U(r)$, with the walls of the pore is given by:

$$U(r) = 0 , \qquad 0 \leq r \leq R - a_s$$
$$U(r) = \infty , \qquad r > R - a_s$$

where r is the distance of the center of the molecule from the axis of the pore, and R is the pore radius (see Fig. 5-5). One now views the pore fluid as divided into two regions: (1) an inner core region of radius $R - a_s$, in which the solute concentration is the same as in bulk

---

essential difference.) Note that osmotic flow is driven by a ΔP within the pores equal to $\sigma RT \, \Delta c_s$ (instead of $RT \, \Delta c_s$ when the solute is impermeant), and that the solute will be acted upon by this flow. (That is why the concentration profile of solute within the pores is not drawn as a straight line.) (B) Volume flow has been stopped by an applied ΔP equal to $\sigma RT \, \Delta c_s$. This figure is analogous to Figure 2-6B, which should be consulted. There is now no pressure gradient within the pores, but equilibrium has not been achieved, because there is still interdiffusion of solute and water between the compartments. (We neglect any small volume movement that may be associated with unbalanced diffusive flow of solute and water.)

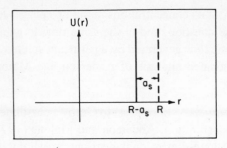

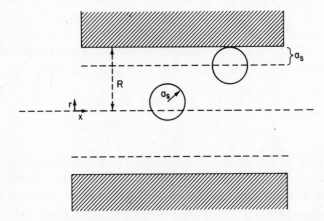

Figure 5-5. A view within a pore of radius R containing solute particles that are hard spheres of radius a_s. The two outer dashed lines are a distance a_s from the pore wall; the center of the solute molecules cannot approach closer than this to the pore wall. *Inset:* The potential energy $U(r)$ of interaction of a hard sphere solute molecule with the pore wall as a function of the distance (r) of its center from the axis of the pore. $U(r)$ is zero until the center of the molecule is a distance a_s from the wall (shown as a dashed line), beyond which it becomes infinite.

74

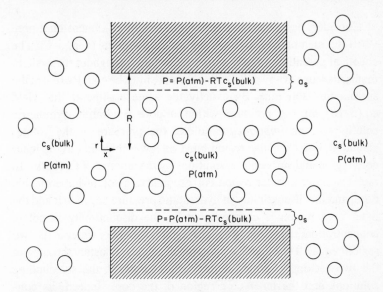

Figure 5-6. The situation within a pore separating symmetric solutions (at atmospheric pressure) of a permeant solute. (As in the previous figure, the solute is a hard sphere of radius a_s.) In the inner core region of the pore ($r < R - a_s$), the solute concentration [c_s(bulk)] and pressure [P(atm)] are the same as in the two solutions. In the outer annular region ($R - a_s < r < R$) the solute concentration is zero, and the pressure is lower than that in the two solutions by the amount RTc_s(bulk).

solution, and (2) a solute-excluded outer annular region (of thickness a_s) of pure solvent,* in which the solute concentration is zero (see Fig. 5-6). The apparent partition coefficient K_s of solute between pore and bulk solution is then given by the ratio of the inner, accessible volume to the total pore volume, or:

$$K_s = \frac{\pi(R - a_s)^2}{\pi R^2} = \left(1 - \frac{a_s}{R}\right)^2 \qquad (5\text{-}21)$$

*The solvent is treated as a continuum; that is, the finite radius of the water molecule is not considered.

Consider, now, the situation within this three-dimensional pore when the membrane separates symmetric solutions (Fig. 5-6). The chemical potential of water must be constant throughout the system; that is, its value at any point within the pore must be equal to that in the surrounding solutions. By exactly the same arguments that yield eq.(5-16) for the pressure within a pore separating symmetric solutions of an *impermeant* solute, the pressure in the solute-excluded, outer annular region must be lower than that in the inner core region and surrounding solutions by the amount RTc_s(bulk). In fact, the outer annular region corresponds to a solute-impermeable pore, and the discontinuity in hydrostatic pressure between it and the inner core region is compensated by the discontinuity in solute concentration, thus making the chemical potential of water in the two regions equal. The chemical potential of solute is constant throughout the two regions to which it has access: namely, the surrounding solutions and the inner core region of the pore. Indeed, its concentration is the same in these two phases.*

We now, as before, extend our analysis from this equilibrium situation to the case in which a Δc_s(bulk) exists between the two solutions. Making the usual assumption of equilibrium conditions at the pore ends, and also assuming that equilibrium conditions continue to exist in the radial direction, we see that the solute concentration gradient along the length of the pore gives rise to a pressure gradient in the opposite direction (Fig. 5-7). Note, however, that this pressure gradient is present only in the outer annular region where solute is excluded. (For the impermeant solute case, solute is excluded from all

---

*In the more general case, where the potential energy of interaction $U(r)$ of solute with pore wall increases continuously as $r \to R$ (rather than becoming infinite for $r > R - a_s$), the solute concentration falls continuously as $r \to R$. The chemical potential of solute is then constant throughout the *entire* system; for dilute solutions this fact is expressed by the solute concentration satisfying the Boltzmann distribution: $c_s(r) = c_s$(bulk) $\exp(-U(r)/RT)$. Instead of the pressure discontinuity between the outer annular and inner core region of the pore (Fig. 5-6), the pressure P changes continuously in the annular region. P as a function of r can then be derived either from the equality of the chemical potential of water, or from the condition of mechanical equilibrium: $dP/dr + c_s(r)\,dU(r)/dr = 0$.

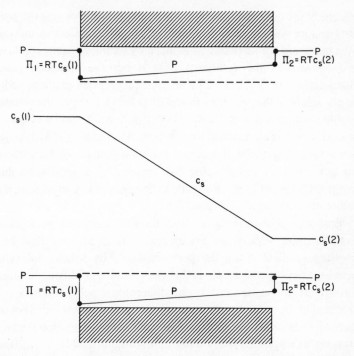

Figure 5-7. Three-dimensional analysis of profiles within the pores of a membrane separating solutions of different concentrations of a permeant solute. Within the inner core region of the pore there is a gradient of solute concentration but no pressure gradient. In the outer annular region, there is no solute present, but there is a pressure difference between the ends equal to $RT \, \Delta c_s$. (If one assumes that equilibrium conditions are maintained in the radial direction, then the pressure gradient in the outer annular region matches the solute concentration gradient in the inner core region, so that at any plane perpendicular to the pore axis, the pressure in the outer annular region is related to the corresponding solute concentration and pressure in the inner core region by the expression $\Delta P = -RTc_s$, where ΔP is the difference in pressure between the outer annular and inner core region.)

regions of the pore, and therefore the pressure gradient acts throughout the pore volume.) Consequently, although osmotic flow induced by a permeant solute (like that induced by an impermeant solute) is still convective, the velocity profile is not the parabolic shape characteristic of flow driven by an external pressure gradient acting on all regions of the pore (i.e., Poiseuillian flow). Instead, the velocity profile resembles that of "plug" flow (Fig. 5-8). A similar plug flow is characteristic of electroosmosis (see Bikerman, 1958). As the solute radius increases, it becomes excluded from a larger fraction of the pore volume, and the plug flow induced by a permeant solute merges into the parabolic, Poiseuillian flow induced by an impermeant solute or an external ΔP (Fig. 5-8).

Both the one-dimensional and three-dimensional analysis of osmotic flow, induced by a permeant solute, attribute flow to a pressure gradient along the pore length.* The salient difference between these two analyses is that the pressure gradient is uniform throughout all regions of the one-dimensional pore, whereas it is confined to the outer annular region (where solute is excluded) of the three-dimensional pore. Consequently, the three-dimensional analysis reveals that (for the leaky solute case) osmotic flow is "plug" flow, basically different from Poiseuillian flow produced by an externally applied hydrostatic pressure difference, whereas the one-dimensional analysis attributes the same flow pattern to both osmotically driven and pressure-driven flow.

There are two sequellae that follow from the more sophisticated, three-dimensional picture of osmotic flow. First, because less flow

---

*This view has been challenged in the one-dimensional analysis carried out by Hill (1983), who claims that although osmosis induced by an impermeant solute is convective flow driven by a pressure gradient, ". . . if solute can enter the pore then osmotic flow is a diffusive phenomenon, and there is no pressure gradient in any part of the pore to which solute has access, even at low concentration due to a repulsive wall field." I cannot undertake here a critique of all of Hill's arguments; suffice it to note that his eq.(8) is a misstatement of the Gibbs-Duhem equation in that it omits the work term involved in bringing solute from solution into the pore against the repulsive wall field. Had Hill included that term, he would have found a pressure drop equal to $RT[c_s \text{(bulk)} - c_s \text{(pore)}]$ at each solution-pore interface [as in eq.(5-17)], and hence the same pressure gradient within the pore as given by Garby's (1957) analysis.

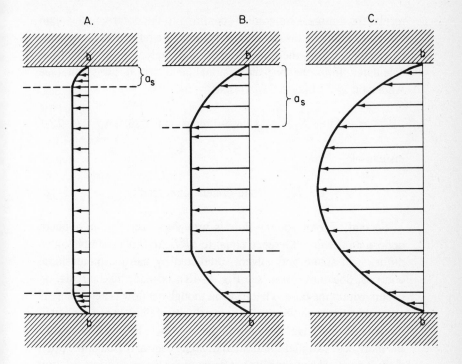

Figure 5-8. Sketch of velocity profiles characteristic of osmotic flow within a pore. The osmotic flow has been generated by a Δc_s of a solute with (A) a radius much smaller than that of the pore, (B) a radius about half that of the pore, and (C) a radius greater than that of the pore (i.e., an impermeant solute). The lengths of the arrows are the distances that liquid particles originally on line *bb* have moved in time *t;* hence, their lengths are proportional to fluid velocity. The velocity is zero at the wall of the pore and reaches a maximum at a distance of a_s from the wall, which it maintains throughout the inner core of the pore, thus giving a "plug flow" character to osmosis. The velocity is constant in the inner core region because there is no pressure gradient there. For osmosis produced by an impermeant solute (C), there is no inner core region, and the plug flow has merged into the parabolic velocity profile characteristic of Poiseuillian flow. This is the same velocity profile that would be generated by an applied hydrostatic pressure difference between the ends of the pore [$P(2) > P(1)$].

results from a pressure gradient confined to the outer annulus than from one acting on the entire pore volume, the reflection coefficient, σ, calculated from the three-dimensional analysis is less than that calculated from the one-dimensional analysis. It turns out (see Anderson and Malone, 1974) that:

$$\sigma = (1 - K_s)^2 \qquad \text{(three-dimensional result)} \qquad (5\text{-}22)*$$

instead of

$$\sigma = 1 - K_s \qquad \text{(one-dimensional result)} \qquad (5\text{-}18)$$

Second, if total volume flow is brought to zero by an applied, opposing pressure difference equal to $\sigma RT \, \Delta c_s$(bulk), simple inter-diffusion of solute and solvent, predicted by the one-dimensional analysis, does not result (see Fig. 5-4B); instead, fluid circulation occurs within the pore. That is, even though net flow is zero, there is *not* mechanical equilibrium, or zero velocity of fluid, at every point in the pore. Rather, near the walls of the pore there is fluid flow toward the more concentrated solution, and, in the central region of the pore, there is an equal and opposite net flow toward the more dilute solution (Fig. 5-9). This pattern of compensating dissipative flows at zero net volume transfer results from the difference in velocity profiles for pressure-driven and osmotically driven flow. A similar flow pattern occurs when electroosmotic flow is brought to zero by an opposing hydrostatic pressure difference [see, for example, Adamson (1967)].

C. UPHILL FLOW OF WATER

An interesting phenomenon can arise if a membrane separates solutions containing solutes with different reflection coefficients. For concreteness and dramatic emphasis, let solutions 1 and 2, separated

---

*This result is probably not exact, since in its derivation, as in the derivation of eq.(5-18), the extra drag on flow created by the presence of solute in the pore was neglected. I comment on this point in the section on irreversible thermodynamics.

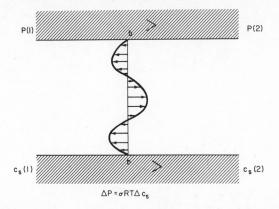

$$\Delta P = \sigma RT \Delta c_s$$

Figure 5-9. Three-dimensional analysis of the velocity profile within a pore when volume flow has been stopped by an applied ΔP equal to $\sigma RT \Delta c_s$. Note that in contrast to the one-dimensional analysis of this situation in Figure 5-4B, there is not simply interdiffusion of water and solute. Instead, there is a balance of convective volume flows: near the walls of the pore, fluid flows from compartment 2 to compartment 1, while in the central region of the pore, an equal flow occurs from compartment 1 to compartment 2.

by the membrane, be $100\,\mathrm{m}M$ urea and $1\,\mathrm{m}M$ dextran, respectively, and let the pores of the membrane be large enough to be freely permeable to urea yet small enough to be totally impermeable to dextran. In other words, $\sigma_{\mathrm{urea}} \approx 0$ and $\sigma_{\mathrm{dextran}} = 1$ (Fig. 5-10A). As usual, we assume that equilibrium conditions prevail at the membrane-solution interfaces and that the two solutions are well stirred and of large volume, so that a steady state exists in which the urea concentration remains essentially $100\,\mathrm{m}M$ in solution 1 and $0\,\mathrm{m}M$ in solution 2 throughout the course of the experiment. In which direction does osmosis occur?

Figure 5-10B illustrates, among other things, the pressure profile within the pores. Because $\sigma_{\mathrm{urea}} \approx 0$, there is no pressure drop at the solution 1-pore interface; in contrast, because $\sigma_{\mathrm{dextran}} = 1$, a pressure drop equal to $RTc_{\mathrm{dextran}} \approx 0.02$ atm occurs at the solution 2-pore

A.

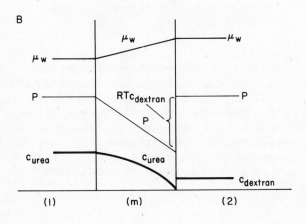

B

C.

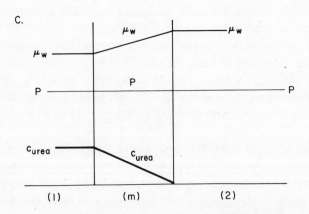

82

interface. Consequently, there is a pressure gradient within the pores that drives water (and urea) from solution 1 to solution 2; that is, osmosis occurs from solution 1 to solution 2. Note that to an observer unaware of the pressure profile within the pores, water is flowing in the "wrong" direction, in that it is flowing from a region of lower chemical potential (100 mM urea) to one of higher chemical potential (1 mM dextran). Even within the pores it is moving against its own chemical potential gradient (see Fig. 5-10B). The reason this can happen is that water transport is dominated by convective flow generated by the dP/dx within the pores. Of course, if dextran were not present in solution 2, there would be no pressure gradient within the pores (since $\sigma_{urea} \approx 0$); water and urea would simply interdiffuse with essentially no osmotic volume transport (Fig. 5-10C). In that instance, water moves in the "right" direction.

Water flow in the "wrong" direction was observed by Meschia and Setnikar (1958) in experiments with a collodian membrane separating urea and dextran solutions comparable to those in our hypothetical example. They anticipated this phenomenon from two previous experiments in which the membrane separated (1) a concentrated urea solution from pure water and (2) a dilute dextran solution from pure water. Given that there was no osmotic flow in the former experiment and substantial osmotic flow in the latter, it was not

---

Figure 5-10. (A) A porous membrane separates a solution of 100 mM urea from a solution of 1 mM dextran. $\sigma_{urea} \approx 0$; $\sigma_{dextran} = 1$. (B) Profiles of pressure, μ_w, and c_{urea} within the pores of the membrane separating the solutions depicted in A. Notice that the pressure gradient is directed toward the dextran solution, and hence osmotic flow occurs in that direction, against the chemical potential gradient of water. The flow of water drags urea along with it. (That is why the concentration profile of urea is shown as a curved line.) (C) Profiles of pressure, μ_w, and c_{urea} within the pores of the membrane separating the solutions depicted in A, except that there is no dextran in compartment 2. Notice that in contrast to B, there is no pressure gradient within the pores, and hence no osmotic flow. There is simply interdiffusion of urea and water.

surprising that in the combined experiment with the membrane separating the urea and dextran solutions, osmosis was toward the dilute dextran solution. It is also not surprising from a theoretical standpoint, once one considers pressure profiles within pores and the influence of σ on the magnitude of the pressure gradient generated by a given Δc_s. We shall return to this experiment at the end of the next section, where we view it from the standpoint of irreversible thermodynamics.

D. IRREVERSIBLE THERMODYNAMICS

It is not my purpose to develop or discuss the foundations of irreversible thermodynamics [see, for example, de Groot (1958)], or to present a general critique of its methodology with respect to transport across membranes (see Katchalsky and Curran, 1965). I intend only to outline the approach taken by this formalism towards osmotic phenomena* and to emphasize the nature of the statements that it can, and cannot, make. My particular purpose is to draw attention to what I believe is a certain confusion generated by irreversible thermodynamics with respect to solute-solvent coupling in osmosis.

Consider once again a membrane separating solutions 1 and 2 of a single solute s. In general, the concentrations of solute may be different in the two solutions, so that there is a Δc_s across the membrane; there may also be a difference in hydrostatic pressure ΔP applied across the membrane. If these gradients are sufficiently small that forces and fluxes can be assumed to be linearly related, it is possible to write for the volume flux J_v and solute flux Φ_s across the membrane (Kedem and Katchalsky, 1958):

$$J_v = L_p(\Delta P - \sigma_{\mathrm{osm}}RT\,\Delta c_s) \tag{5-1}$$

*Our discussion is confined to the "discontinuous" treatment of osmotic phenomena by this formalism.

$$\Phi_s = \omega RT\, \Delta c_s + (1 - \sigma_f)\bar{c}_s J_v \qquad (5\text{-}23)$$

where $\bar{c}_s$ is the mean solute concentration in the membrane ($\approx [c_s(1) + c_s(2)]/2$ and we have distinguished between σ_{osm} and σ_f, the reflection coefficients determined from osmotic and ultrafiltration experiments, respectively. Eq.(5-1) states that volume flow is driven through the membrane by both an applied hydrostatic pressure difference and an osmotic pressure difference; eq.(5-23) describes solute flux through the membrane as the sum of a diffusive component ($\omega RT \Delta c_s$) and a solvent drag component $(1 - \sigma_f)\, \bar{c}_s J_v$.

As we noted earlier, σ_{osm} can be determined either by eq.(5-1a), from the hydrostatic pressure difference that must be applied to stop volume flow:

$$\sigma_{\mathrm{osm}} = \left(\frac{\Delta P}{RT\, \Delta c_s}\right)_{J_v = 0} \qquad (5\text{-}24a)$$

or by eq.(5-1b), from the ratio of the volume flow (in the absence of a ΔP) produced by a given Δc_s of the permeant solute to that produced by the same Δc_s of an impermeant solute; that is,

$$\sigma_{\mathrm{osm}} = -\frac{(J_v)_{\Delta P=0}}{L_p RT\, \Delta c_s} \qquad (5\text{-}24b)$$

On the other hand, if c_s is the same on both sides of the membrane (i.e., $\Delta c_s = 0$ and $\bar{c}_s = c_s$), σ_f can be determined by a filtration experiment, from the ratio of the concentration of solute in the filtrate (Φ_s/J_v) to its concentration in the filtrand (c_s); that is, from eq.(5-23) we have:

$$\sigma_f = 1 - \left(\frac{\Phi_s}{c_s J_v}\right)_{\Delta c_s=0} \qquad (5\text{-}25)$$

CHAPTER 5

Probably the major contribution of irreversible thermodynamics to our understanding of osmotic phenomena is the general result, deriving from the Onsager reciprocal relations, that regardless of the nature of the membrane,

$$\sigma_{osm} = \sigma_f \equiv \sigma \qquad (5\text{-}26)$$

(Staverman, 1951; Kedem and Katchalsky, 1958). Thus, only one σ characterizes a given solute's interaction with a membrane, and its value can be determined independently from either osmotic experiments [eqs.(5-24a and b)] or an ultrafiltration experiment [eq.(5-25)].

The equality of σ_{osm} and σ_f provides an important check on any kinetic model for osmosis induced by a leaky solute.* It is easily shown that $\sigma_{osm} = \sigma_f$ for the solubility-diffusion mechanism of transport described for "oil" membranes.[†] In contrast, σ_{osm} as calculated from eq.(5-22) for a porous membrane is not equal to σ_f calculated for the same type membrane. Anderson and Malone (1974) note this fact but emphasize that they cannot thereby conclude from this that $\sigma_{osm} \neq \sigma_f$, since various approximations made in deriving eq.(5-22), particularly their neglect of the extra viscous drag on flow contributed by the presence of solute in the pore, may account for this inequality. It turns out, in fact, that the difference between the expression for σ_{osm} [eq.(5-22)] and the expression for σ_f, which explicitly considers the solute drag effect in the pore [eq.(63) of Bean (1972)], is small (Anderson and Malone, 1974), thus emphasizing that the major factor contributing to nonzero values of σ is restriction of solute entry into the pore [expressed through K_s in eq.(5-22)].

---

*The *cognoscenti* tend to feel that the Onsager reciprocal relations (at least as applied to membrane phenomena) do not rest on entirely solid theoretical foundations, and that therefore the equality of σ_{osm} and σ_f should be verified for any particular kinetic model. Levitt (1975) has apparently explicitly proved that $\sigma_{osm} = \sigma_f$ for pores to which continuum hydrodynamic theory can be applied.

[†] Multiply eq.(2-17) by $\bar{V}_w$, substitute it and eq.(5-11) into eq.(5-25), and compare this expression (taking c_s small) with eq.(5-10).

Indeed, if $\sigma = 1$ because the solute molecule is too large to enter the pore, it is the *only* factor. Thus, the practice of interpreting the phenomenological coefficients of irreversible thermodynamics in terms of frictional coefficients, to the exclusion of partition coefficients (Kedem and Katchalasky, 1961; Ginsburg and Katchalasky, 1963), can be very misleading. To say that $\sigma \approx 1$ because of a large frictional interaction of solute with the pore wall may be *formally* correct, but the imagery associated with that statement hardly corresponds to reality, if in fact the reason that $\sigma \approx 1$ is because the solute is sterically excluded from entry into the pore.

The reflection coefficient σ is one of the phenomenological coefficients that appears in the irreversible thermodynamic treatment of membrane transport phenomena; L_p and ω of equations (5-1) and (5-23) are the two others. It should be stressed that these phenomenological coefficients are just that—phenomenological. Irreversible thermodynamics has nothing to say about their physical nature. In particular, this discipline makes no statement concerning the nature of L_p, the hydraulic permeability coefficient, or its relation to ω_{THO}; that is, it says nothing about the nature of osmotic transport through a membrane or the value of P_f/P_{d_w}. These are questions whose answers are model-dependent, and they can only be addressed by a specific physical analysis of the model. Of course, once that analysis is performed, the phenomenological coefficients can be explicitly expressed in terms of the parameters of the model. The point is, however, that the phenomenological coefficients themselves are of no help in performing that analysis.

I now close this section with what I consider to be a particularly egregious example of the ability of irreversible thermodynamics, with its phenomenological coefficients, to obfuscate the physics underlying osmotic transport through a porous membrane. Consider again the Meschia and Setnikar (1958) experiment discussed earlier, in which a porous membrane separates 100 mM urea from 1 mM dextran, and $\sigma_{urea} = 0$, whereas $\sigma_{dextran} = 1$. We saw that osmosis proceeds from the urea to the dextran solution, with water moving in the "wrong" direction, from a region of lower to one of higher chemical potential. Mechanistically, of course, this results from the pressure gradient in the pore driving solution in that direction (Fig. 5-10B), but how does the formalism of irreversible thermodynamics interpret this phenomenon? It says that the gain in free energy from water moving *up* its chemical potential gradient is more than compensated by the loss in free energy from urea moving *down* its chemical potential gradient—making the overall free energy of the process negative (in keeping, thank God, with the second law of thermodynamics). Since it is the linking of urea and water transport, expressed by the urea-water coupling coefficients, that accounts for the overall negative free energy of the process, one concludes that water is driven uphill by urea. This has been characterized by one writer as "co-diffusion" of water and solute (Diamond, 1962); that is, solvent flux is coupled to solute flux, and the former is carried uphill by the latter.

This thermodynamic interpretation of the uphill flow of water completely distorts reality; it has everything backwards. *Water,* because of the osmotic effect of dextran, drags urea, not the other way around. If dextran is absent, urea and water simply interdiffuse (Fig. 5-10C); there is no "carrying" of the water by the solute. With dextran present, the resulting pressure gradient within the pore forces the solution within the pore toward the dextran side, and the urea flux is increased as a consequence. Conversely, if the urea is absent, osmotic flow still occurs toward the dextran side; the water does not "need" the urea to cause it to flow in that direction. In fact, to the extent that σ_{urea} is slightly greater than zero, urea actually *retards* the osmotic flow. Becoming enamored with thermodynamic formalism leads to the wrong sign!

Some devotees of irreversible thermodynamics may feel that the 'co-diffusion" interpretation of the Meschia-Setnikar experiment is correct and that consequently our mechanistic description of osmotic flow through porous membranes must be in error. In order to divest this discussion from any possible unease the reader may still have about osmosis, let us take a much simpler situation. Consider a huge sewer pipe with 100 mM urea at one end and pure water at the other.

If a pressure of only a few centimeters of water is applied to the urea end, urea and water will roar through the pipe toward the other end. Now from the thermodynamic view point, water moves uphill against its chemical potential gradient, but the overall free energy change in the system is negative, because of the loss in free energy resulting from urea going down its chemical potential gradient. All of this is exactly comparable to what occurred in the Meschia and Setnikar experiment. I submit, however, that no sane person would claim that urea "carries" water through the pipe and that the reason water flows is because it is "coupled" to the urea flux. Yet this is precisely the description that is given to this event by the formalism of irreversible thermodynamics.

PART

II

LIPID BILAYER MEMBRANES

The water permeability of lipid bilayers has been studied in two different systems. One involves a single planar (or spherical) bilayer separating two macroscopic aqueous compartments; the other consists of a dispersion in water of single walled or multiwalled vesicles, often referred to as liposomes. Since most of the quantitative permeability data come from studies on planar bilayers, I shall focus primarily on that system, with reference, where appropriate, to results obtained on liposomes.

This part of the monograph is divided into three chapters. The first (Chapter 6) discusses the permeability properties in general, and the water permeability in particular, of unmodified lipid bilayers. The other two chapters deal with the water permeability of bilayers that have been modified by the insertion into them of pores (or channels)* formed by the polyene antibiotics nystatin and amphotericin B (Chapter 7) and by the peptide antibiotic gramicidin A (Chapter 8).

*The terms "pore" and "channel" are used synonymously in Chapters 7 and 8, and indeed throughout the entire monograph.

6

The Unmodified Membrane

Since the time of Gorter and Grendel (1925), the idea that the framework of the plasma membrane is a lipid bilayer has played a central role in membrane physiology. This concept received the imprimatur of the biochemical community about 15 years ago (Singer and Nicolson, 1972) and since then has been a major tenet of cell biology, applicable both to the plasma membrane and to the membranes surrounding intracellular organelles. In the early 1960s, Mueller et al. (1963) described for the first time a convenient method of forming between two aqueous phases a planar lipid bilayer whose permeability and electrical properties could be readily studied. Since then, these membranes have been an important research tool in membrane physiology, used for investigating such diverse phenomena as carrier-mediated ion transport, channel gating kinetics, and membrane-vesicle fusion (exocytosis). In most experiments with planar lipid bilayer membranes, the properties of the bilayer *per se* are not of direct interest. Rather, as in the work described in the following two chapters, the bilayer serves as a supporting structure that allows the transport properties of carriers and channels that have been inserted into it to be manifested and thereby investigated, just as the bilayer of the plasma membrane acts as a framework in which the transport proteins embedded in it express their activity; indeed, in recent years, a number of channels from plasma membranes have been functionally incorporated into planar bilayers. In this chapter, however, we consider the general permeability characteristics of the bilayer membrane itself and, in particular, how they relate to the mechanism and magnitude of water permeation across it.

CHAPTER 6

A. THE LIPID BILAYER AS AN OIL MEMBRANE

Qualitative Considerations

To a good first approximation, the permeability properties of lipid bilayer membranes are those expected from a thin layer of hydrocarbon separating two aqueous phases. Thus, their virtual impermeability to small ions such as Na^+, K^+, and Cl^- (as reflected in a very small conductance of $\approx 10^{-8}$ mho/cm^2) (Mueller et al., 1962; Hanai et al., 1964), their very low permeability to polar nonelectrolytes (Vreeman, 1966; Finkelstein, 1976a), and their very high permeability to nonpolar, lipophilic solutes (Orbach and Finkelstein, 1980), are all characteristics expected of a membrane fabricated from a low dielectric constant organic solvent such as hydrocarbon; these characteristics are directly attributable to the solubility (or relative insolubility) of ions and nonelectrolytes in hydrocarbon. In terms of the diagram of a lipid bilayer shown in Figure 6-1, the membrane's permeability properties are those of the hydrocarbon interior created by the hydrophobic tails of the amphipathic lipids (phospholipids and sterols) from which it is formed; the polar head groups of the lipids do not act as a significant permeability barrier, but merely serve to anchor the hydrocarbon region between the two aqueous phases. In other words, if hydrocarbon could be squeezed down between filter paper to a thickness of about 50 Å, one would have a membrane with permeability properties essentially the same as those of lipid bilayers.

This very simple description and explanation of the permeability properties of lipid bilayer membranes has met with some resistance over the years. Part, although not all, of the reluctance to accept this elementary picture may be psychological in origin. It is perhaps somewhat traumatic to have to acknowledge that a physically sophisticated object like a lipid bilayer, which is of molecular dimensions and which has a structure of such potential relevance to that of plasma membranes, has permeability characteristics no more remarkable than those achieved by Overton (1895) and Collander and Bärlund (1933) with ordinary (not even fancy Italian) olive oil.

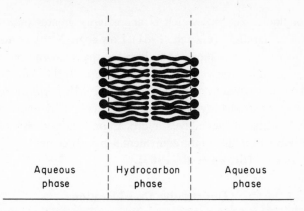

Aqueous phase | Hydrocarbon phase | Aqueous phase

Figure 6-1. Diagram of a lipid bilayer membrane. Phospholipid molecules are schematized by a filled circle representing the polar head group and two wavy lines representing the fatty acid acyl chains. The "liquid" hydrocarbon interior phase of the bilayer, included between the dashed lines, is formed from the acyl chains and is anchored between the two aqueous phases by the polar head groups. Sterols (e.g., cholesterol), not included in the diagram, contribute their rigid, fused hydrocarbon ring structure to the interior phase, thereby making it less fluid [from Finkelstein and Cass (1968)].

(The moral, of course, is that the more interesting permeability properties of plasma membranes, such as ion selectivity, stereoisomer discrimination, voltage-dependent conductances, and active transport, are not inherent characteristics of their lipid bilayers, but rather are associated with the proteinaceous pathways inserted into and through this supporting structure.) It may also be counterintuitive, for many, that the hydrocarbon tails in an apparently ordered bilayer structure should form a phase having properties similar to that of bulk hydrocarbon. This concept, however, is a familiar one in the detergency literature. There, it is well known that the solubility of non-polar solutes in soap and detergent micelles ". . . is strikingly parallel with the solubility in, say, hexadecane," and that "the partial molal volume of the paraffin chains in the micelle is approximately that of liquid paraffin, which is appreciably greater than that of crystalline paraffin" (Garrett, 1961). Low angle X-ray scattering of lipid dispersions has shown that there too the hydrocarbon tails are disordered and much more "liquid" than "solid" in character (Luzzati and Husson, 1962), and recent NMR measurements on unilamellar phosphatidylcholine vesicles suggest that "the segmental microviscosity of the bilayer hydrocarbon region does not differ appreciably from that of the equivalent n-paraffinic liquids of similar chain length" (Brown et al., 1983).

Quantitative Considerations

A more quantitative description of the permeability properties of lipid bilayer membranes than that presented in the preceeding section comes from measurements of nonelectrolyte permeability coefficients, P_d's. Nonelectrolytes for which P_d values can be obtained are, in general, restricted to hydrophilic molecules with hydrocarbon: water partition coefficients (K's)* less than 10^{-3}; the large P_d values of more lipophilic molecules cannot be determined, because of limitations imposed by unstirred layers on the two sides of the bilayer (Holz and Finkelstein, 1970). Exceptions to this limitation are weak acids and weak bases, where the un-ionized form is the permeant species. By working at pH's at which the predominant species is the ionized form (that is, at pH's far above or far below the pK of the acid or base, respectively), it is possible to minimize unstirred layer corrections (Gutknecht and Tosteson, 1973) and thereby determine P_d values for much more lipophilic molecules.

The results from the nonelectrolyte permeability studies by Orbach and Finkelstein (1980) and Walter and Gutknecht (1984) on egg phosphatidylcholine (PC) membranes are presented in Figure 6-2 as a double logarithmic plot of P_d vs. DK, where D is the diffusion coefficient of the nonelectrolyte in water.[†] The data points, which

---

*The hydrocarbon:water partition coefficient K for a solute is defined by eq.(5-5), where the "oil phase" is liquid hydrocarbon (in particular, hexadecane). The subscript "s" to P_d, D, and K in Chapter 5 is omitted in this chapter.

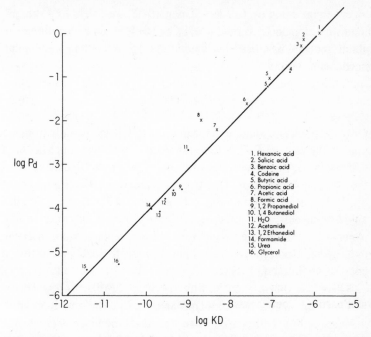

Figure 6-2. Double logarithmic plot of permeability coefficients (P_d) in egg PC membranes vs. *DK,* where *D* is the diffusion coefficient of a molecule in water and *K* is its hexadecane:water partition coefficient. (P_d is in cm/s, *D* is in cm²/s, and *K* is dimensionless.) The data represented by the solid circles (•) are from Orbach and Finkelstein (1980) and those represented by the x's are from Walter and Gutknecht (1984). The line is drawn with a slope of 1. [The value for K_{urea} in this plot is not the one originally determined by Orbach and Finkelstein (1980) and used by them in their graph of their data, but the one recently determined by Walter and Gutknecht (1986), who argue that the original value was spuriously large.]

---

†It would be more appropriate to use the diffusion coefficient of the molecule in hydrocarbon (hexadecane) rather than its diffusion coefficient in water, but since diffusion coefficients in hydrocarbon are not known for most molecules, diffusion coefficients in water (which should be more or less proportional to those in hydrocarbon) are used instead. The issue of the appropriate way to depict diffusion in the interior of a bilayer is discussed later in the chapter.

cover a range of almost 6 orders of magnitude, are well fit by a straight line with a slope of 1, as predicted by eq.(5-4) for an oil membrane in which molecules cross the membrane by a solubility-diffusion mechanism; that is,

$$P_d \propto KD \qquad (6-1)$$

Only the points for water and formic acid differ significantly from the mean—the value of P_d/DK for these molecules being about a factor of 3–5 too large. The mean value of $P_d\delta/DK$, where the thickness (δ) of the hydrophobic interior of the bilayer is taken as 50 Å (Fettiplace et al., 1971), is approximately 0.3 (Finkelstein, 1976a) or 0.6 (Walter and Gutknecht, 1984) for egg PC membranes. In other words, measured P_d values are very close to those predicted from inserting bulk hydrocarbon (hexadecane) partition coefficients and diffusion coefficients into eq.(5-4).

There is no *a priori* reason why the value of $P_d\delta/DK$ should be so close to 1 for all lipid bilayer membranes; in fact, one anticipates that the nature of the hydrocarbon-like interior of the bilayer, and hence the values of P_d, will be lipid dependent, and this turns out to be the case. Parameters that affect the fluidity of the bilayer interior, such as the amount of sterol in the membrane, the length and degree of unsaturation of the phospholipid hydrocarbon tails, and the temperature, all affect P_d in the predicted manner (see Table 6-1). For example, P_d values are over 100-fold smaller in long-chain, saturated, cholesterol-containing sphingomyelin membranes at 15° C, than in short-chain, unsaturated egg PC membranes at 25° C. It is not clear to what extent the effect of membrane fluidity on P_d can be attributed to its effect on partition coefficients of molecules into the hydrophobic interior of the bilayer (K) or to its effect on diffusion coefficients within that region (D), but in either case, the combined effect, as expressed in relation (6-1), is more-or-less independent of the nature of the permeant molecule. That is, *relative* P_d values remain approximately constant, even as *absolute* values change by orders of magnitude with bilayer composition (Table 6–1).

TABLE 6-1. *Permeability Coefficients (P_d's) of Water and Nonelectrolytes Through Lipid Bilayer Membranes of Varying Composition*[a]

Molecule	$10^6 K$	PC	PC/chol	Sph/chol	Sph/chol 14.5°
			$10^7 P_d$ (cm/s)		
1,6-Hexanediol	540	—	2,250	450	—
Isobutyramide	370	—	1,980	118	—
n-Butyramide	360	—	3,000	288	51
1,4-Butanediol	43	2,600	200	30	—
H_2O	42	22,000	5,730	810	210
Acetamide	21	1,650	196	21	—
Formamide	7.9	1,030	269	25	—
Urea	0.28[b]	40	6.1	—	—

[a]The membranes were formed from egg phosphatidylcholine (PC), egg phosphatidylcholine plus cholesterol (PC/chol), and bovine sphingomyelin plus cholesterol (Sph/chol). All values were obtained at 25°C, except those in the column labeled 14.5°. K is the hexadecane:water partition coefficient of the molecule. [From Finkelstein (1976a).]
[b]From Walter and Gutknecht (1986).

CHAPTER 6

Although the permeability characteristics of lipid bilayers are explained to a good first approximation by viewing their hydrophobic interior as a simple, liquid hydrocarbon phase into which molecules partition and diffuse as in bulk hydrocarbon, there are measurable departures from this elementary picture. The permeability coefficients for water and other small molecules, such as formamide (Poznansky et al., 1976; Finkelstein, 1976a) and formic acid (Walter and Gutknecht, 1984), are anomalously high, whereas the P_d's for large molecules, such as codeine, and branched molecules, such as isobutyramide, are abnormally small, particularly in tight (i.e., less fluid) bilayers (Finkelstein, 1980). These differences can be attributed to the interior of the bilayer not being truly isotropic, liquid hydrocarbon; the hydrocarbon chains of the lipids do, after all, have some degree of order within the bilayer, the more so the less fluid the phase. If straight chain compounds enter and diffuse across the membrane in an extended conformation with their long axis parallel to the hydrocarbon chains of the lipid, this would explain why, for a given membrane, P_d/DK is constant in going from the two-carbon chain acetamide to the four-carbon chain butyramide or even to the six-carbon chain 1,6-hexanediol, but is smaller for the branched chain isobutyramide and even smaller for the bulky codeine; that is, the cross-sectional areas of the straight chain compounds are all approximately the same, whereas those of isobutyramide and codeine are larger. The anomalously large values of P_d/KD for small molecules such as water and formic acid could similarly be attributed to their small cross sectional areas.

It should be stressed that the considerations in the preceeding paragraph are relatively minor variations on the theme that the permeability of molecules through lipid bilayers is governed by Overton's rule as expressed in relation (6-1). The *relative* permeabilities of molecules are determined primarily by the product of their partition coefficients and free diffusion coefficients in bulk hydrocarbon. The *absolute values* of the permeability coefficients are scaled by the lipid composition. Comparing, for example, acetamide and 1,4-butanediol, the ratio of their P_d's is essentially the same in an egg PC membrane as in a sphingomyelin-cholesterol membrane, although the values of the P_d's are almost 100-fold smaller in sphingomyelin-cholesterol membranes than in egg PC membranes (Table 6-1).

Of the two terms in relation (6-1) that determine the relative permeability coefficients of solutes in a bilayer, the partition coefficient (K) is by far the more important. Figure 6-2 would not be significantly altered if the abscissa were simply K rather than DK. This is because the partition coefficients into hydrocarbon of two small molecules (MW < 500) can easily differ by many orders of magnitude, whereas the diffusion coefficients, being roughly proportional to the cube root of molecular weight, will differ by no more than a factor of 3 or so. The claim has been made that the interior of a bilayer should be viewed as a soft polymer network, like rubber, rather than as a fluid, like liquid hydrocarbon (Lieb and Stein, 1971). Apparently diffusion coefficients within soft polymer networks show a steep, non-Stokesian dependence on molecular volume (V) quite different from the $V^{1/3}$ dependence in liquids, and recent experiments by Walter and Gutknecht (1986) suggest that the anomalously high permeability coefficients mentioned previously for very small molecules like water and formic acid are best explained by viewing transverse diffusion through the bilayer interior as non-Stokesian in nature. The nature of the diffusion process within the bilayer is a very interesting physicochemical problem, but in focusing on it, one should not lose sight of the overwhelming importance of the partition coefficients in determining the relative magnitudes of the experimentally measurable and physiologically relevant permeability coefficients $(P_d$'s).

Most measurements of permeability coefficients on planar bilayer membranes have been obtained on membranes made by variations of the technique originally described by Mueller et al. (1963). In particular, the membrane is formed from a solution of lipid dissolved in decane. It is known, however, that a considerable amount of decane remains dissolved in the bilayer (Fettiplace et al., 1971), thus raising the possibility that the hydrocarbon-like property of the bilayer emphasized in this section is not an intrinsic property of the bilayer structure, but results instead from the presence of decane within it. It appears on several counts, however, that this is not a serious cause for concern. First, membranes formed from lipids dissolved in hexadecane, rather than in decane, retain much less hydrocarbon solvent (Fettiplace et al., 1971), yet P_d values in PC membranes formed this way are only slightly smaller than in PC membranes formed with decane as the solvent (Orbach and Finkelstein, 1980). Second, for "hydrocarbon-free" bacterial phosphatidylethanolamine (PE) bilayer membranes formed by the technique developed by Montal and Mueller (1972), in which the amount of hydrocarbon solvent in the bilayer is minimal (Benz et al., 1975), $P_{d_{acetic\ acid}}$ is actually somewhat larger than in decane-containing bacterial PE membranes (Walter and Gutknecht, 1984). Finally, relation (6-1) remains valid in cholesterol-containing membranes, even though cholesterol substantially reduces the amount of decane retained in the bilayer (Haydon et al., 1977). All of these data, therefore, indicate that the hydrocarbon-like permeability properties of lipid bilayer membranes are intrinsic to the bilayer structure and do *not* depend on the presence of hydrocarbon solvent in the membranes.

B. THE WATER PERMEABILITY OF LIPID BILAYER MEMBRANES*

The preceding section discussed the general permeability properties of lipid bilayer membranes and described transport as occurring across them by a solubility-diffusion mechanism characteristic of an oil membrane—the "oil" in this case being the hydrocarbon-like interior of the bilayer. For these membranes, P_d values are proportional to the product of the hydrocarbon:water partition coefficient and diffusion coefficient of the molecules (relation 6-1), with the former quantity being the dominant term. In all these respects, water

---

*Additional details, particularly regarding liposomes, can be found in the review by Fettiplace and Haydon (1980).

itself is not an exception, and its mechanism of transport is the same as that of any other "solute," although, as we noted, the value of P_d/DK for water in a given membrane is somewhat larger than the mean value for other solutes.

Consistent with the oil membrane picture of a lipid bilayer, P_f is found to be equal to P_{d_w} [eq.(3-4)], in those membranes where it is possible to make the determination (Cass and Finkelstein, 1967; Holz and Finkelstein, 1970). I pointed out in Chapter 3 that the presence of unstirred aqueous layers can lead to spuriously low values of P_{d_w} and hence to overestimations of P_f/P_{d_w}; this turns out to be a problem for determinations of P_f/P_{d_w} on planar lipid bilayer membranes (Cass and Finkelstein, 1967; Everitt et al., 1969; Holz and Finkelstein, 1970). Correct values for P_{d_w} are obtained, however, on "tight" membranes, which have water permeabilities considerably smaller than those of the unstirred layers in series with them, and then it is found that eq.(3-4), $P_f/P_{d_w} = 1$, is obeyed. Because of the unstirred layer problem in measuring P_{d_w}, most reported values of bilayer water permeability coefficients are of P_f's, determined in osmotic experiments, that are not significantly affected by unstirred layers for the reasons discussed in Chapter 3.

The temperature dependence of P_f is also consistent with the lipid bilayer being equivalent to a thin layer of liquid hydrocarbon (in so far as its permeability characteristics are concerned). The Q_{10} for the partition coefficient (solubility) of water in liquid hydrocarbon (hexadecane) is 1.5 and that for its diffusion coefficient in hexadecane is 1.2 (Schatzberg, 1965). This leads to the prediction, from relation (6-1), that the Q_{10} for P_{d_w} (and hence for P_f as well) should be 1.8, and this is approximately what is found experimentally for membranes formed from a single lipid such as PC (Graziani and Livne, 1972; Fettiplace, 1978). [For cholesterol-containing membranes, somewhat larger Q_{10}'s are obtained (but see Graziani and Livne (1972)), because the amount of cholesterol in the bilayer is a function of temperature (Redwood and Haydon, 1969).] Conceivably, this agreement of experiment with theory results from the presence of residual hydrocarbon solvent in planar bilayer membranes, and the measured

Q_{10}'s do not reflect the true temperature dependence of P_f for solvent-free lipid bilayers; this is not likely, however, as comparable Q_{10} values are obtained on liposomes, which contain no dissolved hydrocarbon (Reeves and Dowben, 1970; Blok et al., 1977).

The most interesting physiological aspect of planar lipid bilayer water permeability (which comes up again in the discussion of plasma membrane water permeability in Chapter 9) is the absolute values of P_f. Depending on lipid chain length, degree of chain unsaturation, cholesterol content, etc.—in other words, the same parameters discussed earlier that affect the membrane's fluidity, and therefore its permeability to any molecule—P_f (or P_{d_w}) values have been obtained which are as low as 2×10^{-5}* and as high as 1×10^{-2} cm/s (Huang and Thompson, 1966; Finkelstein, 1976a), a span of almost three orders of magnitude. These numbers encompass almost the entire range of values reported for cell membranes. Thus, one need look no further than their lipid bilayer framework to account for the *magnitude* of P_f for cell membranes; as we shall see in Part III, however, there are other issues with respect to the water premeability of cell membranes that require explication besides the magnitude of P_f. The same arguments, given in the previous section, for believing that the residual hydrocarbon in planar bilayers does not significantly affect P_d values for solutes are equally valid for water. In addition, the value obtained for P_f on unilamellar egg PC vesicles (Reeves and Dowben, 1970) is in general agreement with the values obtained on planar bilayers formed from egg PC (see Fettiplace and Haydon, 1980).

In the following two chapters, the water permeability induced in planar lipid bilayer membranes by certain pore-forming molecules is discussed. As we have just seen, the unmodified membrane has a substantial water permeability of its own that will act in parallel with any inserted pathway. It is this factor that has thus far limited water permeability measurements of pores in lipid bilayers to those formed by the polyene antibiotics nystatin and amphotericin B and by the

---

*Values about 5-fold lower have been obtained on lipid vesicles below their phase transition (Lawaczek, 1979).

peptide antibiotic gramicidin A. Only with these molecules have sufficient numbers of pores been inserted into the bilayer to enable the water movement through them to be recorded above the background of water flux through the rest of the bilayer. Even so, the lipid composition had to be carefully chosen so that the bilayer had as low a water permeability as possible consistent with the ability of the antibiotics to insert very large numbers of pores into it. For all of the data discussed in the next two chapters on the water permeability induced by the pores, the background water permeability of the unmodified bilayer has already been subtracted, under the assumption that it remains constant, independent of antibiotic concentration or the number of pores in the membrane. The justification for this assumption is given in the discussion of the results obtained with these antibiotics.

C. A NOTE ON UNSTIRRED LAYERS

A matter of practical concern in measuring *true* values of P_d for solutes and water in lipid bilayer membranes is to determine the magnitude of the unstirred layer correction that must be applied to the experimentally *observed* values. This requires determining the unstirred layer thickness δ associated with the membrane ($= \delta_1 + \delta_2$, the sum of the unstirred layer thicknesses on the two sides of the membrane), and a simple way of doing this is to measure the experimentally observed diffusion permeability coefficient, $(P_d)_{obs}$, for a lipophilic solute. Equation (3-8) can be generalized to any molecule; that is,

$$(P_d)_{obs} = \frac{1}{1 + P_d(\delta/D)} P_d \qquad (6\text{-}2)$$

where P_d is the true membrane permeability coefficient of the molecule and D is its diffusion coefficient in water. For a lipophilic molecule such that P_d is much greater than D/δ (the unstirred layer permeability coefficient of the molecule), eq.(6-2) reduces to

$$(P_d)_{\text{obs}} \approx \frac{D}{\delta} \qquad\qquad (6\text{-}2a)$$

and δ is directly obtained from $(P_d)_{\text{obs}}$.

A convenient lipophilic molecule that can be used to determined δ is butanol. With the usual experimental arrangement and stirring conditions for making permeability measurements on planar lipid membranes, it is found that $\delta \approx 100 \ \mu m$ (Holz and Finkelstein, 1970). Two observations indicate that butanol is indeed so permeable through lipid bilayers that eq.(6-2a) is satisfied. First, if the viscosity (η) of the medium is increased with glucose, $(P_d)_{\text{obs}}$ for butanol is decreased proportionately (Holz and Finkelstein, 1970). This is just what one expects from eq.(6-2a), since $D \ \alpha \ 1/\eta$. [It is also possible that the increase in viscosity increases δ, which also will contribute to the reduction in $(P_d)_{\text{obs}}$.] Second, in contrast to results given in Table 6-1 for hydrophilic solutes, in which P_d's change by orders of magnitude as a function of membrane composition, $(P_d)_{\text{obs}}$ for butanol remains constant (Finkelstein, 1976a). This is precisely what should occur for a molecule so permeant that $(P_d)_{\text{obs}}$ is independent of its actual membrane permeability coefficient (P_d) and is entirely determined by the unstirred layers.

Once δ has been determined, eq.(6-2) can be used to calculate the true value of the permeability coefficient of a molecule (P_d) from the observed value. All of the P_d's in Table 6-1 and the P_d's described in the next two chapters for antibiotic-modified membranes have been corrected for unstirred layers. In general, these corrections are either trivial or small. It is a wise policy for the experimentalist to disregard data for which the unstirred layer correction affects the value of P_d by a factor of two or more (that is, if $P_d \geq D/\delta$). As can be seen from eq.(6-2), when P_d is greater than D/δ, small errors in the determination of δ can lead to very large errors in the calculated value of P_d.

7

Nystatin and Amphotericin B

The polyene antibiotics are a large group of antifungal agents produced by the genus *Streptomyces*. They are all characterized by a large polyhydroxylic lactone ring of 23–37 carbon atoms, with 4–7 conjugated double bonds in the ring. Many polyenes have in addition a carboxyl group and an amino sugar glycosidically linked to the ring. Their antifungal activity derives from their ability to bind to the plasma membrane of fungi and make it "leaky," thereby killing the cell. This action on cell membranes is restricted to those containing sterol, thus making them toxic to fungi, which contain ergosterol as the sterol in their plasma membrane, but ineffective on bacteria, whose membranes are sterol free. They also act on the cholesterol-containing membranes of animal cells, which consequently limits their efficacy as therapeutic agents for internal use. (For reviews of the chemistry, biosynthesis, and general properties of these antibiotics, see Hammond, 1977; Martin, 1977; Medoff and Kobayashi, 1980.)

The three polyenes whose actions on cell membranes and lipid bilayers have been most extensively studied are filipin and the clinically useful antibiotics nystatin and amphotericin B (Fig. 7-1). All three molecules in micromolar quantities cause cell lysis, but although addition of sucrose to a normal saline medium protects cells from the lytic action of the latter two (e.g., Cass and Dalmark, 1973), it does not protect them from the lytic action of filipin. These and related results on cells and liposomes suggest that nystatin- and amphotericin B-induced lysis is a colloid osmotic lysis resulting from the creation by these antibiotics in sterol-containing plasma membranes of pores large enough to allow Na^+, K^+, Cl^-, and 5-carbon

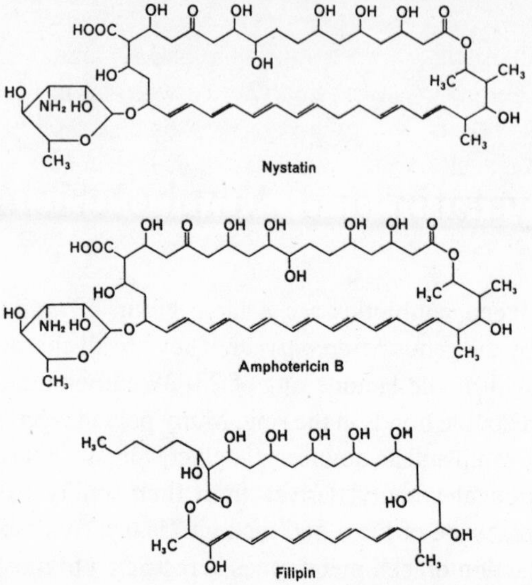

Figure 7-1. Structural formulae of nystatin, amphotericin B, and filipin [as given in Medoff and Kobayaski (1980)].

sugars to permeate, but small enough to exclude glucose and sucrose, whereas filipin-induced lysis results from massive lesions that permit the non-specific efflux of macromolecular cytoplasmic components such as glucose-6-phosphate dehydrogenase (e.g., de Kruijff et al., 1974).

The action of these antibiotics on planar lipid bilayer membranes is consistent with the results obtained on plasma membranes and liposomes. Treatment of sterol-containing planar bilayers with filipin simply produces membrane instability and breakage (Andreoli and Monahan, 1968), whereas treatment with nystatin and amphotericin B induces much more subtle and interesting phenomena that are the subject of this chapter. The effects of nystatin and amphotericin B on planar bilayer membranes are very similar; unless otherwise indicated, descriptions in this chapter of the action of either one can be assumed to apply to both. Chemically, the two molecules are almost identical (Fig. 7-1). Both consist of a large polyhydroxyl lactone ring, to which is attached a carboxyl group and an amino sugar (mucosamine). The only chemical difference between the molecules is in the location of two of the hydroxyl radicals and in the number of conjugated double bonds in the lactone ring; amphotericin B is a heptaene, whereas one of those double bonds is hydrogenated in nystatin, thereby yielding a tetraene and a diene separated by a methylene bridge.

A. TWO-SIDED ACTION*

Ion Permeability

When added in micromolar amounts to both sides of a sterol-containing planar bilayer separating symmetric NaCl or KCl solutions, nystatin and amphotericin B can increase membrane conductance by more than seven orders of magnitude, from the unmodified membrane conductance of 10^{-8} mho/cm^2 to values of over 10^{-1} mho/cm^2. This antibiotic-induced conductance is proportional to a large power of the antibiotic concentration, ranging from 4–12 depending on membrane lipid composition, and results from an increase in membrane permeability to both univalent anions and cations, with the permeability to anions substantially greater than that to cations (Andreoli and Monahan, 1968; Cass et al., 1970). The selectivity among univalent anions appears to be dependent on ionic size, with a permeability sequence of $Cl^- > F^- > CH_3SO_3^- > HOCH_2CH_2SO_3^-$ (Finkelstein and Cass, 1968). The preference of nystatin- and amphotericin B-treated membranes for anions over cations must be attributed to the hydroxyl groups of the ring and not to the amino sugar attached to it, since the same preference is displayed by membranes treated with the N-acetyl derivatives of the molecules

---

*In a later section we shall see that nystatin or amphotericin B addition to only one side of a planar bilayer membrane produces an effect that is somewhat different from that produced by their two-sided addition, and that a very interesting relationship exists between their one-sided and two-sided actions.

(Dennis et al., 1970; Finkelstein and Holz, 1973). There are a number of interesting aspects of the ion permeability induced in lipid bilayer membranes by nystatin and amphotericin B [see particularly, Brutyan and Yermishkin (1983)], and some of them will be mentioned subsequently in this and the following chapter; however, since this monograph does not delve into the physics of ion permeation through membranes, we turn from this aspect of nystatin and amphotericin B action to consider their effect on water and nonelectrolyte permeability.

Water and Nonelectrolyte Permeability

Parallel with the increase in conductance (ion permeability) induced by these antibiotics is a proportional increase in water and small nonelectrolyte permeability (Holz and Finkelstein, 1970); antibiotic-induced P_d's for water and small solutes increase linearly with membrane conductance, as illustrated in Figures 7-2 and 7-3 for water and urea. For purposes of comparison, it is convenient to normalize P_d values to those of a membrane with any arbitrary conductance, as is done in Table 7-1. As is evident from the table, P_d values decrease with increasing molecular size; for nystatin-treated membranes:

$$P_{d_w} : P_{d_{urea}} : P_{d_{ethylene\ glycol}} : P_{d_{glycerol}} = 1 : 0.08 : 0.04 : 0.01 \quad (7\text{-}1)$$

The same *relative* selectivity is found for amphotericin B-treated membranes; at the same antibiotic-induced conductance, however, the *absolute* values of P_d are approximately a factor of 2 smaller for amphotericin B-treated membranes (see Table 7-1)—a point commented upon below.

These data in and of themselves strongly suggest that nystatin and amphotericin B form pores in lipid bilayer membranes; discrimination among nonelectrolytes on the basis of molecular size (i.e., "sieving" of molecules), as reflected in eq.(7–1), is the classic criterion used by physiologists to argue that transport across a given membrane occurs

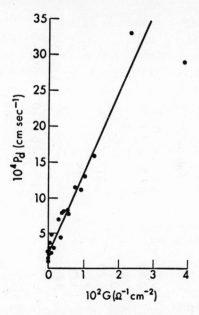

Figure 7-2. P_{d_w} as a function of conductance (G) for nystatin-treated membranes. [From Holz and Finkelstein (1970).]

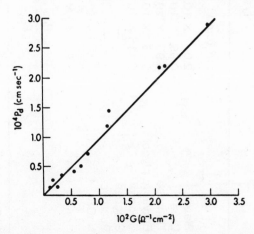

Figure 7-3. $P_{d_{urea}}$ as a function of conductance (G) for nystatin-treated membranes. [From Holz and Finkelstein (1970).]

TABLE 7-1. *Permeability of Nystatin- or Amphotericin B-Treated Lipid Bilayer Membranes*

	Nystatin		Amphotericin B	
	P_d(cm/s)	σ	P_d(cm/s)	σ
Water	12.0×10^{-4}	0	6.0×10^{-4}	0
Urea	0.95×10^{-4}	0.55	0.68×10^{-4}	0.57
Thiourea	0.95×10^{-4}	—	—	—
Ethylene glycol	0.45×10^{-4}	0.67	—	—
Glycerol	0.115×10^{-4}	0.78	0.075×10^{-4}	—
Glucose	—	1.0	—	1.0
Sucrose	—	1.0	—	—
NaCl	—	1.0	—	1.0

$$P_f = 40 \times 10^{-4} \text{ cm/s} \qquad P_f = 18 \times 10^{-4} \text{ cm/s}$$

$$\frac{P_f}{P_{d_w}} = 3.3 \qquad \frac{P_f}{P_{d_w}} = 3.0$$

All P_d's and P_f's are values normalized for membranes of conductance 10^{-2} mho/cm^2 in 0.1 M NaCl. [From Holz and Finkelstein (1970).]

through pores. In addition, two further observations lend over-whelming support to this interpretation. First, along with the increase in P_{d_w} produced by nystatin and amphotericin B is a corresponding increase in P_f (Fig. 7–4); at any given conductance, the value of P_f is greater than that of P_{d_w}. In particular (see Table 7–1):

$$\frac{P_f}{P_{d_w}} \approx 3 \qquad \text{(nystatin- or amphotericin B-treated membrane)} \quad (7\text{–}2)$$

As discussed in Chapter 3, the implication of $P_f/P_{d_w} > 1$ is that water molecules traverse the membrane through a water-rich region, and the simplest realization of this is a water-filled pore.

The second observation indicating a porous permeability pathway is that the reflection coefficients, σ's, of small hydrophilic solutes are

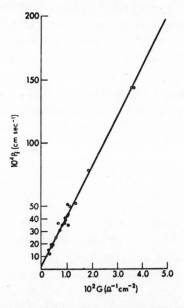

Figure 7-4. P_f as a function of conductance (G) for nystatin-treated membranes. Osmotic gradients produced with NaCl ($\bullet$) or glucose ($\circ$). [From Holz and Finkelstein (1980).]

113

less than 1. The trend in σ values parallels that in P_d values; that is, the smaller the σ the larger the P_d, with σ increasing from 0.55 for urea to 1 for glucose and sucrose (Table 7–1). Comparing the nonunity σ values in Table 7-1 with the corresponding P_d values, we see that the σ's are much lower than $1 - P_{d_s}\bar{V}_s/P_f\bar{V}_w$, the value predicted for a solute that does not interact with water as it crosses the membrane [see eq.(5-10)]. As discussed in Chapter 5, this finding is also anticipated for solutes that cross a membrane through aqueous pores.

Thus, on the basis of the macroscopic permeability characteristics of nystatin- and amphotericin B-treated membranes—namely, the graded permeability to small polar nonelectrolytes, as reflected in both P_d and σ values, and the fact that $P_f/P_{d_w} > 1$—it is clear that these antibiotics form aqueous pores in lipid bilayer membranes. The direct demonstration of single-channel activity in membranes treated with low concentrations of these antibiotics (Ermishkin et al., 1976; 1977) removes any lingering doubt that they form pores. The single-channel conductance of the amphotericin B pore is about twice that of the nystatin pore (Ermishkin et al., 1976), which is consistent with the macroscopic observation that at a given conductance, P_d (and P_f) values for amphotericin B-treated membranes are about half those for nystatin-treated membranes (Table 7-1); that is, at a given conductance, an amphotericin B-treated membrane contains half as many pores as a nystatin-treated membrane.

Size of the Pores

A simple estimate of pore size can be made from the finding that the σ for glycerol is approximately 0.8 and that for glucose and sucrose is 1 (Table 7-1). In other words, the nystatin and amphotericin B pore is quite tight to glycerol and essentially impermeable to glucose and sucrose. On this basis a pore radius of approximately 4 Å can be assigned. It should be recognized, however, that σ determinations are not so accurate that one can readily distinguish between a value of 0.99, which means the solute is permeant, albeit poorly so, and a value of 1.0 for a totally impermeant solute. It is therefore encouraging that two other estimates of pore radius are in agreement with this value of 4 Å.

One independent estimate of pore radius (r) is obtained by substituting the value of 3 for P_f/P_{d_w} [eq.(7-2)] into either eq.(3-6), (3-6a), (3-6b), or (3-6c). [For eqs.(3-6b) or (3-6c), the radius of the water molecule (a_w) is taken to be 1.5 Å.] Values for r of 6 Å, 5 Å, 5 Å, and 4 Å are found, respectively, from eqs.(3-6) through (3-6c). The lack of any theoretical justification for applying these equations to pores of molecular dimensions was discussed in Chapter 3; nevertheless, it is comforting that values for r obtained from them correspond to the value estimated from the presumed cut-off in permeability to molecules larger than glucose. A second estimate of pore radius can be made from restricted diffusion theory based on the Renkin equation (Renkin, 1954), which is eq.(3-5a) written for any molecule instead of just for water.* Despite the theoretical limitations of this equation discussed in Chapter 3, it is reassuring that when the P_d values of Table 7-1 are inserted into it and their ratios taken, a pore radius of approximately 4 Å is found. Thus, by three separate criteria—cut-off size for permeant molecules, the magnitude of P_f/P_{d_w}, and relative values of P_d—a radius of about 4 Å is obtained for the nystatin- and amphotericin B-created pore.

Model of the Pore

A space-filling molecular model (CPK) of nystatin or amphotericin B reveals several striking features of these molecules that, in conjunction with their effects described in the previous sections on membrane conductance, water permeability, and nonelectrolyte permeability, immediately suggest a molecular structure for the pore. In Figure 7-5 both faces of amphotericin B are shown, (A) and (B), along with a CPK model of a phospholipid (phosphatidylcholine) for

---

*An analogous equation can be written for σ's instead of for P_d's (Renkin, 1954). Calculations based on σ values, however, are intrinsically less accurate than those based on P_d's, and for reasons discussed by Holz and Finkelstein (1970), σ values in Table 7-1 may be underestimated, particularly those for urea and ethylene glycol. For these reasons, σ's were not used to estimate pore radius.

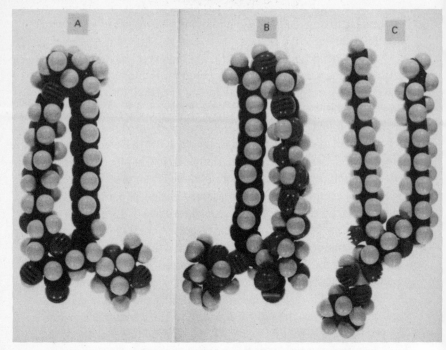

Figure 7-5. Molecular model (CPK) of amphotericin B. In (A) the complete hydrophobic face of amphotericin B is seen; in (B) the molecule has been rotated 180° about its long axis to reveal the opposite face with its many hydroxyl groups. Note that the molecule consists of two chains: a polyene chain [seen on the right in (A) and on the left in (B)] and an amphipathic chain. The hydrophobic and hydrophilic faces of the amphipathic chain are seen in (A) and (B), respectively. At the bottom of the figures are the polar amino sugar and carboxyl group; at the top is a single hydroxyl group [seen most clearly in (B)]. In (C), a CPK model of phosphatidylcholine is shown for comparison with amphotericin B. [From Finkelstein and Holz (1973).]

comparison (C). Amphotericin B, with a length essentially the same as that of a phospholipid, can be characterized as a perverted phospholipid in which its polar end consists of a carboxyl group and an amino sugar, rather than a phosphate plus another group, and its two chains are rigidly fixed with respect to one another; the chain containing the polyene chromophore is completely hydrophobic (like the acyl chains of a phospholipid), whereas the chain containing the large number of –OH groups has a hydrophilic face and a hydrophobic face. An additional conspicuous feature of the molecule is the single –OH group at its nonpolar end.

The dependence of amphotericin B-induced conductance and nonelectrolyte permeability on a large power of the antibiotic concentration indicates that the pore is multimeric.* The most obvious orientation of the amphotericin B molecule, by analogy to that of the phospholipids, is with its polar aminosugar-carboxyl end anchoring it to the aqueous phase and its two linked chains extending into the bilayer parallel to the phospholipid acyl chains. Although this orientation is energetically unfavorable for a *single* amphotericin B molecule, since it places the –OH face of its amphipathic chain among the hydrocarbon tails of the phospholipids, a number of them (8–10) with this alignment can be packed together to form a cylinder as shown in Figure 7-6, with a water-filled interior lined by the hydroxyl groups, and a completely nonpolar exterior, thereby making the structure compatible with the hydrophobic milieu of the bilayer. Two such cylinders can be held together in the center of the bilayer by a ring of hydrogen-bonded –OH groups formed from the single hydroxyl group at the nonpolar end of each molecule, thus accounting for the antibiotic's greater effectiveness in increasing membrane permeability when added to *both* sides of a membrane (Cass et al., 1970). In addition, cholesterol or ergosterol packs snugly in the wedge between each pair of amphotericin B molecules (see Fig. 7-6), thus providing a natural explanation of the sterol requirement for polyene activity.

This is the presently accepted model for the nystatin and amphotericin B pore, proposed independently by Finkelstein and Holz

---

*Since the power varies with membrane lipid composition, yet the pore's intrinsic permeability characteristics are lipid invariant, it is clear that the molecularity of the pore cannot be equated with the value of the power. The relationship between them can be complicated and is beyond the scope of this monograph; a discussion of this issue is given by Finkelstein and Holz (1973).

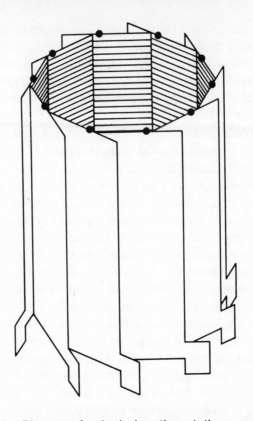

Figure 7-6. Diagram of a single-length nystatin or amphotericin B pore. Each polyene molecule is schematized as a plane with a protuberance and a solid black dot. The protuberance represents the amino sugar, and the solid black dot represents the lone hydroxyl group at the nonpolar end of the molecule; the shaded portion of each plane, partially visible in the center of the pore, represents the polyhydroxyl face of the amphipathic chain. Note that the interior of the pore is polar, whereas the exterior is completely nonpolar; the cleft between each pair of polyene molecules in the exterior of the pore can accommodate a sterol molecule. The double-length pore is formed from two single-length pores hydrogen bonded in the middle of the membrane through the ring of hydroxyl groups (black dots). [From Finkelstein and Holz (1973).]

(1973) and de Kruijff and Demel (1974). The internal radius of a pore constructed from 16 polyene molecules (8 molecules in a cylinder and two cylinders forming the pore) is approximately 4 Å, a size consistent with the nonelectrolyte and water permeability characteristics of nystatin- and amphotericin B-treated membranes discussed earlier. Although there are no independent physical data on polyene-treated membranes (e.g., spectroscopic studies) to confirm this structure, it is nonetheless very appealing. It is the most natural way for these molecules to generate a pore (of appropriate internal radius) with a hydrophobic exterior and a polar (hydroxyl) interior, thus enabling a water-filled, high dielectric constant pathway to span a region of low dielectric constant hydrocarbon. As noted above, the model depicted in Figure 7-6 is that of a "half-pore," the complete pore being formed by the union, through hydrogen bonds in the center of the bilayer, of two such structures, each spanning half of the bilayer, and thus accounting for the two-sided action of nystatin and amphotericin B. Interestingly, however, these molecules also have a one-sided action closely related to their two-sided activity, which, as we shall see, is explained by essentially this same model.

B. ONE-SIDED ACTION

When added to only one side of sterol-containing planar lipid bilayer membranes, nystatin and amphotericin B can, on appropriate membranes (see below), produce increases in membrane conductance as large as those generated by their two-sided action. Conductance is again proportional to a large power of the antibiotic concentration, and membrane permeability to both univalent cations and anions is increased, but in contrast to the two-sided effect, the permeability in this instance is much greater for cations than for anions (Marty and Finkelstein, 1975). Comparable conductances are usually achieved with much smaller concentrations of antibiotic added to both sides than to one side of the membrane (particularly in cholesterol-containing, as opposed to ergosterol-containing, membranes), but it is clear that the one-sided effect is not produced by a contaminant in nystatin and amphotericin B preparations and that the same molecule is responsible for both the one-sided and the two-sided actions (Kleinberg and Finkelstein, 1984). How can the same molecule induce cation selectivity when added to one side of a membrane and anion selectivity when added to both sides?

The answer to this question, proposed by Marty and Finkelstein (1975), is that the "half-pore" shown in Figure 7-6 is also capable of spanning a bilayer as a functional pore; that is, nystatin and amphotericin B can form both double-length and single-length pores. This proposal might appear far-fetched, as it would not seem possible that the thickness of the hydrocarbon interior of a bilayer could simultaneously equal one and two lengths of a nystatin or amphotericin B molecule. The proposal, however, is predicated on the assumption that bilayer thickness is not immutable, and that acyl chains can fully stretch in the region of a double-length pore to accommodate it and compress in the region of a single-length pore (Fig. 7-7). The

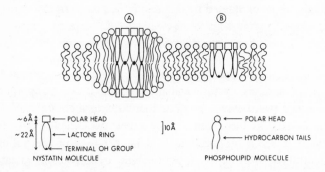

Figure 7-7. Diagram of a double-length (A) and single-length (B) nystatin (or amphotericin B) pore sitting in the same lipid bilayer membrane. A region of unmodified bilayer is depicted between A and B. Note that the acyl chains of the phospholipids can lengthen considerably beyond their extent in the unmodified bilayer region to accommodate the long, double-length pore. [From Kleinberg and Finkelstein (1984).]

thickness of the hydrophobic core of an unperturbed bilayer is about half the length of two fully extended acyl chains (Small, 1967; Lecuyer and Dervichian, 1969), and thus comparable to the long dimension of the nystatin or amphotericin B lactone ring. In other words, there is a play of about a factor of 2 in the thickness of the bilayer interior; the smallest (compressed-chain) thickness, which corresponds to that of an unperturbed bilayer, is compatible with a single-length pore, and the longest (fully extended-chain) thickness accommodates a double-length pore.

Two recent experimental findings support the thesis that the pores formed by the one-sided and two-sided actions of nystatin and amphotericin B have basically the same architecture, and that the only significant structural difference is in their length. One is that the relative values of the nonelectrolyte permeability coefficients for the two pores are identical; in particular, $P_{d_{urea}}/P_{d_{glycerol}} = 8$ in both cases (Kleinberg and Finkelstein, 1984). Since the pores markedly discriminate between two molecules (urea and glycerol) whose radii differ only by about 1 Å, the interaction of the pore walls with these molecules must be substantial; therefore, comparatively small differences in pore radii would be reflected in large differences in the relative permeability coefficients of these molecules. The finding that $P_{d_{urea}}/P_{d_{glycerol}}$ is the same for both pores is therefore convincing evidence that their internal radii are identical.

The second finding that supports the argument depicted in Figure 7-7 is the dependence of the one-sided action of the polyenes on bilayer thickness. If single-length pores exist, their formation should be critically dependent on this parameter, and that is what is found experimentally with monoglyceride/ergosterol membranes (Kleinberg and Finkelstein, 1984). Membranes formed from monoglycerides with acyl chains of 14, 16, or 18 carbon atoms are very sensitive to nystatin's one-sided action, whereas those formed from monoglycerides with acyl chains of 20, 22, or 24 carbon atoms are totally refractory to its one-sided action. Apparently membranes formed from monoglycerides with acyl chains of 20 carbons or longer are too thick to accommodate the pore induced by the one-sided action of nystatin; the pore cannot span the membrane. A comparison of the length of nystatin's lactone ring with the thickness of the bilayer interior (as determined from capacitance measurements) makes this interpretation compelling (Figure 7-8).

There is no obvious reason why the single-length pore is cation selective, whereas the double-length one is anion selective. As previously noted, the ion selectivity appears to be determined by the hydroxyl groups lining the pore lumen, but *a priori* their dipole moments could orient to make it either negative or positive, thereby favoring cations or anions, respectively. If we assume the former is the preferred orientation in both pores, the cation preference of the single-length pore is explained. The anion preference for the double-length pore might then arise from an anion well or cation barrier created in the middle of the pore by the extra hydrogens from the hydroxyls that hydrogen bond to hold the two half pores together.

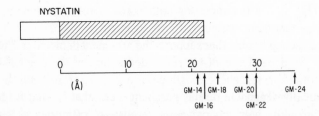

Figure 7-8. Schematic comparison of the length of the hydrophobic portion (hatched) of nystatin (or amphotericin B), as measured on a CPK molecular model, with the thickness of the hydrophobic interior of monoglyceride/ergosterol membranes, as experimentally determined from membrane capacitance measurements. GM-14, GM-16, . . . refer to monoglyceride/ergosterol membranes formed from monoglycerides having acyl chain lengths of 14, 16, . . . carbon atoms, respectively. [From Kleinberg and Finkelstein (1984).]

C. WATER PERMEABILITY PER PORE

Theory and Assumptions

By combining the single-channel conductance measurements with macroscopic permeability data, one can calculate the permeability coefficient *per pore* for water or any other nonelectrolyte. Thus for any molecule we can write [see eq.(4-12)]:

$$p_d \equiv \frac{P_d A}{n} \tag{7-3}$$

and in addition for water there is,

$$p_f \equiv \frac{P_f A}{n} \tag{4-1}$$

where n is the number of pores in the membrane. If membrane conductance (G) is measured along with P_d and P_f, as is generally the practice, n can be obtained from:

$$n = \frac{G}{g} \tag{7-4}$$

where g is the single-channel conductance, and substituting this into eqs.(7-3) and (4-1), the single pore permeability coefficients can be calculated from eqs.(7-5) and (7-6):*

$$p_d = P_d A \frac{g}{G} \tag{7-5}$$

---

*The area (A) of the membrane does not even have to be known for this calculation, since P_d and P_f are obtained by dividing the flux (Φ), which is the experimentally measured quantity, by A [see eqs.(5-3) and (2-12)]. Hence, the values of $P_d A$ and $P_f A$ are independent of membrane area.

$$p_f = P_f A \frac{g}{G} \tag{7-6}$$

Although the recording of single-channel conductances for many different types of channels in a wide variety of cells is now commonplace, the measurement of single-channel water and nonelectrolyte permeability coefficients is unique to the channels formed by nystatin and amphotericin B and to that formed by gramicidin A (see Chapter 8).[†] The values of these permeability coefficients, particularly those for water, therefore take on a special significance, as they provide a means for estimating water and nonelectrolyte permeability coefficients of other channels of molecular dimension.

An important unstated assumption is made in using the preceding equations to calculate single-pore permeability coefficients. Permeability measurements are, by necessity, performed on membranes containing many pores ($\approx 10^9$–10^{11} pores/cm^2), as fluxes through fewer pores are experimentally undetectable either because they are intrinsically too small or because they are small compared to the background flux through the unmodified bilayer. In contrast, single-channel conductances (g) are measured on membranes containing one or very few pores. The use of eqs.(7-5) and (7-6) to compute single-pore permeability coefficients assumes that g is the same in membranes containing $\approx 10^{10}$ pores/cm^2 as in membranes containing only a few pores. There is, however, no guarantee that this assumption is valid. For example, electrostatic interactions among ions entering or traversing neighboring pores may modify single-channel conductances, thereby introducing an unspecified error in the permeability coefficient values calculated from eqs.(7-5) and (7-6). It would be more accurate to use values of g determined by noise analysis on membranes containing 10^9–10^{11} pores/cm^2 in calculating permeability coefficients from eqs.(7-5) and (7-6), but in practice mean membrane conductances are so large and fluctuations about the mean so small,

[†] Single-channel nonelectrolyte (but not water) permeability coefficients have also been determined for the channel formed by the cytolytic toxin from the sea anemone *Stoichactis helianthus* (Varanda and Finkelstein, 1980).

that this has not proved possible. For the present we can only acknowledge this potential source of error in calculating single-pore permeability coefficients and hope that it does not seriously distort their values.

Values

The water permeability coefficients per pore, p_{d_w} and p_f, for single-length and double-length nystatin and amphotericin B pores are given in Table 7-2, along with theoretical values of these quantities for right circular cylindrical pores of radius 4 Å and length either 25 Å (single-length pore) or 50 Å (double-length pore). The agreement of the experimental values with the predictions from macroscopic theory is rather striking.

Consider first p_f. Its value of 1.0×10^{-13} cm³/s for the single-length pore is within an order of magnitude of that predicted from Poiseuille's law alone [eq.(2-21)], and within a factor of two of that predicted from Poiseuille's law corrected for the finite radius of the water molecule [eq. (2-21c)]. Similar good agreement with these equations is found for the double-length pore. [p_f for the double-length amphotericin B pore is half that of the single-length nystatin pore, as is to be expected. p_f for the double-length nystatin pore, however, is one seventh, rather than one half, that of the single-length pore. This discrepancy of about a factor of three may reflect an error in the single-channel conductance value used to calculate p_f for the double-length pore; the possibility for such an error was discussed earlier in conjunction with the validity of eq.(7-6).]

The experimental value of p_{d_w} is somewhat further off from the value predicted by Fick's law [eq.(3-5)] than p_f is from Poiseuille's law, but agrees within a factor of two with the value predicted from Fick's law corrected for the finite radius of the water molecule [eq.(3-5a)]. [p_{d_w} for the double-length nystatin pore is, like p_f, about a factor of three smaller than expected.] The ratio of p_f to p_{d_w} as calculated from eqs.(2-21c) and (3-5a) is about 3.5, in agreement with the values of 3.0 and 3.3 found experimentally.

TABLE 7-2. *Water Permeability Coefficients Per Pore (p_f and p_{d_w}) for Single-Length and Double-Length Nystatin and Amphotericin B Pores*

	Experimental	Poiseuille's Law [eq.(2-21)]	Modified Poiseuille's Law [eq.(2-21c)]
	$10^{14}\,p_f(cm^3/s)$	$10^{14}\,p_f(cm^3/s)$	$10^{14}\,p_f(cm^3/s)$
Nystatin (single-length)	10	62	18
Nystatin (double-length)	1.5	31	9
Amphotericin B (double-length)	4.5	31	9

	Experimental	Fick's Law [eq.(3-5)]	Modified Fick's Law [eq.(3-5a)]
	$10^{14}\,p_{d_w}$ (cm³/s)	$10^{14}\,p_{d_w}$ (cm³/s)	$10^{14}\,p_{d_w}$ (cm³/s)
Nystatin (single-length)	3.0	40	5
Nystatin (double-length)	0.45	20	2.5
Amphotericin B (double-length)	1.5	20	2.5

Experimental values of p_f and p_{d_w} for nystatin pores were calculated by multiplying the $p_{d_{urea}}$ values in 0.1 M KCl (Kleinberg and Finkelstein, 1984) by 42 and 12.6, respectively—the ratios of P_f and P_{d_w} to $P_{d_{urea}}$ (see Table 7-1). Experimental values of p_f and p_{d_w} for amphotericin B were calculated from the P_f and P_{d_w} values of Holz and Finkelstein (1970) [see Table 7-1] and a single-channel conductance in 0.1 M NaCl of 2.5×10^{-13} mho [extrapolated from the data in Fig. 3 of Ermishkin et al. (1977)]. The theoretical values are those of a right circular cylinder of radius 4 Å and length 25 Å (single-length) or 50 Å (double-length). a_w in eqs.(2-21c) and (3-5a) was taken as 1.5 Å.

CHAPTER 7

In Chapters 2 and 3, I emphasized that eqs.(2-21) to (2-21c) and eqs. (3-5) and (3-5a) are derived from macroscopic hydrodynamic theory and that their application to pores of molecular dimensions lacked theoretical justification. It is, therefore, quite remarkable that the values of p_f and p_{d_w} predicted by these equations for a pore of 4 Å radius should agree so well with those obtained experimentally for the nystatin and amphotericin B pores (whose radius of 4 Å was independently inferred from nonelectrolyte sieving experiments). As noted in Chapter 2, Levitt's (1973) molecular dynamics calculations for a pore of about 4 Å radius lead to the same conclusion with respect to p_f. It appears that these equations retain a validity in a domain where the original assumptions underlying their derivation are no longer valid. This is not the only instance of macroscopic hydrodynamic equations remaining valid at the molecular level; a most notable example is the success of the Stokes-Einstein equation in calculating molecular radii from diffusion constants.

D. AREA OF MEMBRANE OCCUPIED BY PORES

The P_f and P_{d_w} values in Table 7–1 are those attributable to the polyene pores and were obtained by subtracting the background water permeability coefficients of the unmodified membrane from the experimentally measured values. Implicit in this procedure is the assumption that water flux through the bilayer proper remains the same, independent of the polyene concentration or the number of pores in the membrane. This assumption is justified by the experimental results themselves, for P_{d_w} and P_f are linear functions of membrane conductance (that is, the number of polyene pores), and at zero conductance (no pores) the lines go through the points obtained on unmodified membranes (see Figs. 7-2 and 7-4). If, in addition to forming pores, nystatin and amphotericin B significantly altered the water permeability of the bilayer proper, these results would not have been obtained.

It is not surprising that the bilayer was largely unaffected by nystatin and amphotericin B in the experiments in which P_f and P_{d_w}

were obtained, when we consider the area occupied by the pores in those membranes. The membrane conductances were around 10^{-2} mho/cm^2 (see Figs. 7-2 and 7-4), and single-channel conductances are about 10^{-13} mho under the conditions of the experiments. Thus, there were about 10^{11} pores per cm^2 of membrane area. Since the cross sectional area of a nystatin or amphotericin B pore is about 50 Å^2 (πr^2, $r = 4$ Å), only about 0.05% of the membrane area is occupied by pores even at this high density. (The percent is larger, but still very small, if the outer, rather than the internal, pore radius is used in the calculation.) It is therefore not surprising that the background water permeability of the lipid bilayer is unaffected by large numbers of nystatin or amphotericin B pores embedded in it.

8

Gramicidin A

Gramicidin A, an antibiotic produced by *Bacillus brevis,* is a linear polypeptide consisting of 15 hydrophobic amino acids in an alternating L-D sequence (Fig. 8-1); both of its ends are blocked— the N-terminal (the head) by a formyl group and the C-terminal (the tail) by an ethanolamine group.* It was one of the first antibiotics to be isolated (Hotchkiss and Dubos, 1940) and was, in fact, discovered during an explicit search by Dubos for bactericidal agents produced by soil microorganisms (Dubos, 1939). In submicromolar concentrations it can increase the conductance of planar lipid bilayer membranes by more than seven orders of magnitude, and, as we shall see, it does this through the formation of cation-selective pores. The pore-forming properties of gramicidin A have been extensively studied, and probably more is known about transport through this pore than through any other. [Ironically, despite the distinguished history of gramicidin A as an antibiotic and the detailed knowledge that exists about the pore (which accounts for the molecule's antibiotic activity), probably its major biological function is to inhibit RNA polymerase during bacterial sporulation, an activity that is unrelated to its pore-forming ability (Paulus et al., 1979).] Although the ion-conducting and ion-selectivity properties of the pore have

*Most gramicidin A preparations are a mixture of valine gramicidin A (the major component, shown in Fig. 8-1) and isoleucine gramicidin A; these molecules differ only in their N-terminal amino acid (Sarges and Witkop, 1965a;b). Commercial preparations of gramicidin contain, in addition to gramicidin A, lesser amounts of valine and isoleucine gramicidin B and C (Glickson et al., 1972); these differ from gramicidin A in having at position 11 an L-phenylalanine and an L-tyrosine, respectively, instead of an L-tryptophan (Sarges and Witkop, 1965c). The actions of all these molecules on lipid bilayer membranes are very similar.

CHO − L-Val − Gly − L-Ala − D-Leu

L-Ala − D-Val − L-Val − D-Val − L-Trp − D-Leu

L-Trp − D-Leu − L-Trp − D-Leu − L-Trp − NHCH₂CH₂OH

Figure 8-1. Structure of valine-gramicidin A. Each horizontal row of amino acids corresponds to approximately one helical turn of Urry's β_6-helical model of the gramicidin A pore. The two diagonal lines represent peptide bonds connecting the three helical turns. The helix is stabilized by intramolecular hydrogen bonds connecting the amino acids obliquely; that is, each amino acid in the pore interior is hydrogen-bonded to the two amino acids immediately adjacent to its vertical neighbors. [From Finkelstein and Andersen (1981).]

received the most attention, water movement through the pore has not been totally neglected, particularly as it relates to ion transport. The main focus of this chapter will, of course, be on water permeation, but to place this in context, some general background information on the pore and its ion-permeability properties is first provided. The reader interested in more details (of which there is no dearth) about both the theoretical and experimental aspects of ion movement through the pore can consult a number of recent papers (e.g., Andersen, 1983a,b; Hladky and Haydon, 1984).

A. PORE STRUCTURE

The conductance induced in lipid bilayer membranes by gramicidin A reflects an increased membrane permeability to univalent cations; the gramicidin A-modified membrane is, in fact, ideally permeable to small univalent cations (Myers and Haydon, 1972; Urban et al., 1980). This macroscopic conductance and univalent cation permeability of gramicidin A-treated membranes results from the formation in the membrane of cation-selective pores with a unit conductance of about 5×10^{-12} mho in 0.1 M NaCl (Urban et al., 1980). The pore formed by gramicidin A was one of the first two pores ever discovered and studied at the single-channel level (Hladky and Haydon, 1970; 1972) [the first being EIM (Bean et al., 1969; Ehrenstein et al., 1970)] and was the first for which the chemical formula of the pore-forming molecule was known. Consequently, a great deal is known both about its properties and about its structure.

It is now generally accepted that the pore is a hydrogen-bonded, head-to-head dimer of a β_6-helix (originally called a $\Pi_{L,D}^6$ helix), as first proposed by Urry (1971; Urry et al., 1975). The evidence supporting this model is substantial and will not be reviewed here (e.g., Bamberg et al., 1977; Weinstein et al., 1979). The salient features of the pore structure are evident in its CPK molecular model shown in Figure 8-2. As with the nystatin and amphotericin B pore, its exterior surface is completely hydrophobic (Fig. 8-2A), thereby making it compatible with the bilayer interior, whereas its lumen (Fig. 8-2B) is lined by polar groups (here, by the NH and $C = 0$ groups of the peptide backbone) to provide a high dielectric constant milieu for water and ions. Thus, the gramicidin A and polyene pores have the same general topological scheme. (Indeed, on *a priori* grounds, any pore spanning a bilayer must have this general scheme; that is, it must have a hydrophobic exterior and a polar interior.) There is, however, an important structural distinction between these two types of pores. The lumen of the polyene pore is formed from subunits that are arranged like staves of a barrel, whereas the lumen of the gramicidin A pore passes through a helix formed from a single molecule. (There are, of course, two helices joined head to head, but each helix has a lumen formed from one molecule, not one built from multiple units.) The proposed structures for the acetylcholine channel (e.g., Guy, 1984; Finer-Moore and Stroud, 1984) are of the polyene class, not so much because they are formed from subunits, as that the α-helices forming the pore are grouped in a barrel stave-like arrangement. This type of arrangement is also proposed for the seven helices of a single bacteriorhodopsin molecule forming the photon-driven proton pump (Henderson and Unwin, 1975).

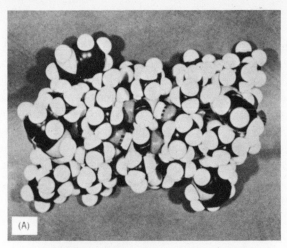

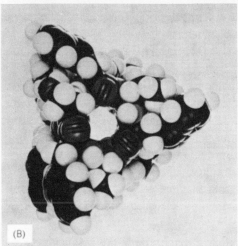

Figure 8-2. Molecular model (CPK) of the gramicidin A pore in the left-handed β_6-helical form. In A the completely hydrophobic exterior surface is seen; in B the pore has been rotated 90° about a transverse axis to reveal its cylindrical lumen of 2 Å-radius lined by the polar groups (NH and C=O) in the peptide backbone. The transmembrane pore is a dimer of two gramicidin A molecules joined by three hydrogen bonds at the formyl ends (see middle of A); the pore entrance is at the ethanolamine ends (seen in B). [From Urry (1972).]

The most notable feature of the model of the gramicidin A pore is the lumen size of approximately 2 Å radius. This mandates, on steric grounds alone, that solutes even as small as urea are excluded from the pore, and in fact gramicidin A-treated membranes are impermeable to urea (Finkelstein, 1974). In addition, and of more interest, water and ion transport through the pore must occur by a single-file process. The single-file aspect of water movement and water-ion interaction within the pore forms a major theme in this chapter, but before turning to it, let us briefly review the general ion-permeability characteristics of the pore.

B. ION PERMEABILITY

The relative permeability of the pore to alkali cations, based on either single-channel conductances (Hladky and Haydon, 1972) or biionic potentials (Myers and Haydon, 1972) in $0.1\,M$ salt solutions, is $P_{Cs^+} > P_{Rb^+} > P_{K^+} > P_{Na^+} > P_{Li^+}$. Discrimination among these cations by the pore is not extraordinary; e.g., $P_{K^+}/P_{Na^+} \approx 3$. H^+ permeability is anomalously high (Hladky and Haydon, 1972; Myers and Haydon, 1972) in the same sense that its mobility in free solution is. The latter results from a Grotthus mechanism of transport (that is, H^+ hops from one water molecule to another rather than H_3O^+ diffusing as an entity); the large proton conductance of the gramicidin A pore suggests that proton transport occurs there by a similar mechanism—a point discussed later in connection with single-file transport.

Translocation of ions through the gramicidin A pore is usually described in terms of the energy profile for an ion. Two examples of such profiles are shown in Figure 8-3. In both, there are two energy minima (or sites), one at each end of the pore, separated from each other (and the aqueous phases) by energy barriers which the ion must traverse when moving into, through, and out of the pore. [The justification for the size and location of the energy barriers and minima is beyond the scope of this discussion and remains a subject of active research (see, for example, Levitt, 1978a,b; Urry et al.,

1982a,b; Andersen, 1983a,b; Pullman and Etchebest, 1983; Etchebest and Pullman, 1984; Mackay et al., 1984).] It is generally believed that the pore can be simultaneously occupied by at most two ions, one in each well (e.g., Levitt, 1978a; Urban and Hladky, 1979; Hladky and Haydon, 1984). The binding constants for these sites are a function of the ion species. Thus, it has been argued that even at the highest NaCl concentrations, pores contain only one sodium ion (Finkelstein and Andersen, 1981), whereas at 0.1 M CsCl a significant fraction are occupied by two cesium ions (Hladky and Haydon, 1984). The two energy profiles shown in Figure 8-3 differ in the magnitude of the central electrostatic energy barrier. In Figure 8-3A, this barrier is substantial and governs the rate of ion movement between the wells; in Figure 8-3B, on the other hand, it is negligible, and ion movement between the wells is governed by local interactions between the ion and the pore walls, and by single-file coupling to water movement. This latter aspect will be discussed in a subsequent section.

C. WATER PERMEABILITY

Number of Water Molecules in the Pore

As already noted, the 2 Å radius of the gramicidin A pore restricts the water phase within it to a single row, as in Figure 4-1, and demands that water movement occur by a single-file process. Thus, the concept of single-file transport discussed *theoretically* in Chapter 4 is *physically* realized with the gramicidin A pore, and therefore experimental results obtained on gramicidin A-treated membranes can be interpreted in terms of that theory. Alternatively, one can view the gramicidin A results as a test of single-file theory, particularly with respect to the predicted value of P_f/P_{d_w}, which is still controversial(see Chapter 4). The opportunity provided by the gramicidin A pore to experimentally study single-file transport (of both water and ions) is a major reason for its having been so extensively investigated.

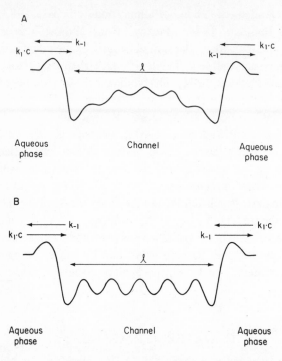

Figure 8-3. Possible energy profiles for ion translocation through a gramicidin A pore. (The profiles have been drawn for a single ion moving through the pore and do not depict the changes which occur with multiple-ion occupancy.) The energy levels of the minima near the ends of the pore are drawn below that of the ion in free solution, because univalent cations are favored in the pore relative to free solution (Myers and Haydon, 1972). (A) Energy profile with a central electrostatic energy barrier. The short-range local interactions between the ion and the pore wall and the electrostatic repulsion of the ion from the pore interior combine to form two energy minima (two sites), one at each end of the pore, separated from each other (and from the aqueous phases) by energy barriers which the ion must traverse when moving through the pore [see Levitt (1978b) and Andersen and Procopio (1980)]. (B) Energy profile without a central electrostatic energy barrier. The electrostatic barrier may be sufficiently small that ion movement through the pore becomes almost exclusively limited by local interactions between the ion and the pore wall, as well as by

The water permeability coefficients (P_f and P_{d_w}) of gramicidin A-treated membranes are linear functions of the gramicidin A-induced conductance and extrapolate at zero conductance (that is, at zero gramicidin A-induced conductance) to the value for the unmodified membrane (Fig. 8-4). From the slopes of the lines in Figure 8-4, one obtains (Rosenberg and Finkelstein, 1978b):

$$\frac{P_f}{P_{d_w}} = 5.3 \qquad \text{(gramicidin A-treated membrane)} \qquad (8\text{-}1)$$

By the same sorts of arguments presented in the preceding chapter for the polyene pores, it is clear that the expression in eq.(8-1) for P_f/P_{d_w} is a property of the gramicidin A pore, and that even at the high membrane conductances (i.e., large numbers of pores) at which the water permeability coefficients were determined, only a small fraction of the membrane area is occupied by pores. It is striking that the value of P_f/P_{d_w} for the 2 Å-radius gramicidin A pore (≈ 5) is greater than that of the 4 Å-radius nystatin and amphotericin B pore (≈ 3). This experimentally supports the theoretical arguments advanced in Chapter 4 that the value of P_f/P_{d_w} for a single-file pore is not a function of pore radius, but is instead a function of N, the number of water molecules in the pore. [Equation (8-1) also conflicts with Manning's (1975) contention that $P_f/P_{d_w} = 1$ in a single-file pore.] In fact, if we accept Levitt's analysis that:

$$\frac{P_f}{P_{d_w}} = N \qquad (4\text{-}17)$$

---

single-file, flux-coupling to water. The energy profile will again show the same features as illustrated in A, but the physical meaning of the central barrier has changed considerably. The figures illustrate the various rate constants: k_1 is the rate constant for association of an ion with an empty pore (k_1c is the rate of ion entry); k_{-1} is the exit (or dissociation) rate constant of an ion from a singly-occupied pore; l is the rate constant for ion translocation through the pore interior. [From Finkelstein and Andersen (1981).]

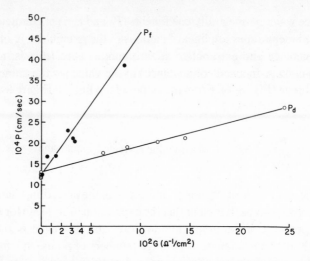

Figure 8-4. P_f (●) and P_{d_w} (○) of gramicidin A-treated membranes as a function of membrane conductance (G). Note that for the unmodified membrane ($G \approx 0$), $P_f = P_{d_w}$. *The slopes of the P_f and P_{d_w} lines are 34.2 × 10⁻³ and 6.5 × 10⁻³ cm s⁻¹/Ω⁻¹ cm⁻²*, respectively. [From Rosenberg and Finkelstein (1978b).]

then,

$$N \approx 5 \qquad \text{(gramicidin A pore, based on } P_f/P_{d_w}) \qquad (8\text{-}2)$$

Water Permeability Per Pore

The slopes of the P_f and P_{d_w} lines in Figure 8–4 are 34.2×10^{-3} and 6.5×10^{-3} cm s⁻¹/Ω⁻¹ cm⁻², respectively. Those experiments were performed on PE membranes in 0.01 M NaCl solutions.* Taking the single-channel conductance in this solution as 2.8×10^{-13} mho [based on Andersen's (1978) value of 2.8×10^{-12} mho for PE membranes in 0.1 M NaCl], Rosenberg and Finkelstein (1978b) obtained single-pore water permeability coefficients of 9.6×10^{-15}

---

*For technical reasons (see Rosenberg and Finkelstein, 1978b), the solutions also contained 0.1 M choline chloride.

cm^3/s and 1.8×10^{-15} cm^3/s for p_f and p_{d_w}, respectively. These calculations of water permeability coefficients for a single gramicidin A pore from macroscopic permeability coefficients and single-channel conductance measurements are, of course, subject to the same caveat discussed in the preceeding chapter with respect to the validity of eqs.(7-5) and (7-6). From osmotic experiments on gramicidin A-treated glycerol monoolein membranes and the single-channel conductance measurements of Neher et al. (1978), Dani and Levitt (1981a) obtained 6×10^{-14} cm^3/s for p_f, a value 6-fold larger than that determined by Rosenberg and Finkelstein (1978b). The cause of this difference is not clear. Given the difficulties in determining P_f and accurately measuring large membrane conductances (see Rosenberg and Finkelstein, 1978b; Dani and Levitt, 1981a), the discrepancy most likely results from one or more methodological errors.

We saw in the preceeding chapter that the values of p_f for the 4 Å-radius nystatin and amphotericin B pores were in surprising agreement with those predicted from macroscopic hydrodynamic theory. It is even more remarkable to find this same concordance of experiment with macroscopic theory for the 2 Å-radius gramicidin A pore. A naive application of Poiseuille's law [eq.(2-21)] to a pore of radius 2 Å and length 25 Å gives a value for p_f of 3×10^{-14} cm^3/s, which falls between the two reported experimental values for the gramicidin A pore of 1×10^{-14} and 6×10^{-14} cm^3/s (Rosenberg and Finkelstein, 1978b; Dani and Levitt, 1981a). Of course, movement of water molecules through the gramicidin A pore can hardly be characterized as Poiseuillian flow; one shudders to think what a parabolic velocity profile in a single-file pore could mean.

D. ION-WATER INTERACTION

In any aqueous pore through which both water and ions move, there will in principle be mutual interactions between the two fluxes, thereby generating electrokinetic effects (e.g., streaming potentials and electroosmosis). Electrokinetics is one of the most arcane subjects in physical chemistry and would therefore seem an inappropriate topic in a monograph that has avoided discussing ionic physics even in the context of osmotic *equilibrium,* let alone in relation to transport phenomena. We saw in Chapter 4, however, that ion-water interactions take on a particularly interesting, and tractable, form in single-file transport; in fact, the number of water molecules (N) in a single-file pore can be obtained from electroosmotic and streaming potential measurements [eqs.(4-19) and (4-20]. In this section we shall consider this and related issues pertaining to ion-water interaction in the single-file gramicidin A pore.

Number of Water Molecules in the Pore

Rosenberg and Finkelstein (1978a) and Levitt et al. (1978) determined N, the number of water molecules in a gramicidin A pore, from streaming potential and electroosmotic measurements on gramicide A-treated membranes. The values obtained by these two groups are now more or less in accord, although some detailed differences still remain. Several technical difficulties arise in measuring streaming potentials and electroosmotic flow, and, as discussed by Rosenberg and Finkelstein (1978a), are probably the source of any disagreements between the reported values of N. I shall focus here on the results of Rosenberg and Finkelstein (because of the well-known principle enunciated by Einstein that "einiger Dreck stinkt nicht"), noting where appropriate the findings by Levitt et al. (1978) that are at variance with them.

In 0.01 and 0.1 M NaCl, KCl, and CsCl solutions, streaming potentials across gramicidin A-treated membranes are 3.0 mV per osmolal difference in nonelectrolyte concentration. From eq.(4-20), this means that $N = 6.5$; that is, there are approximately 7 water molecules in the gramicidin A pore:

$$N \approx 7 \quad \text{(gramicidin A pore, based on streaming potentials)} \quad (8-3)$$

[Levitt et al. (1978) originally reported approximately 12 water molecules in the pore; more recently, Levitt (1984) has modified this number to 9 or 7, depending on whether the measurements are made in low concentrations of NaCl or KCl, respectively.] In 1 M solutions of these salts, streaming potentials are only 2.35 mV per osmolal difference in nonelectrolyte concentration, thereby reducing N to 5.1. [Levitt (1984) claims that N remains 9 even at the highest NaCl concentrations, but falls to 5 at high KCl concentrations.] There are at least two explanations for the smaller value of N at high salt concentrations. One is that there are fewer water molecules in the pore, because the osmotic stress withdraws one or two water molecules from it. (See Chapter 2 and Fig. 2-6C for comments on the state of tension within a pore at high impermeant solute concentrations.) The other is that N in eq.(4-20) represents the number of water molecules transferred through the pore *per ion* (Finkelstein and Rosenberg, 1979). If a pore never contains more than one ion, N equals the number of water molecules in the pore. On the other hand, at high salt concentrations where the pore may contain at times two ions, N can be less than this number; in fact, for a multiion-occupied pore, N is the average number of water molecules between the ions.

Interestingly, no measurable streaming potentials are generated by osmotic gradients across gramicidin A-treated membranes separating 0.01 M HCl solutions (Levitt et al., 1978; Rosenberg and Finkelstein, 1978a). This is consistent with the Grotthus mechanism of proton movement discussed earlier in connection with the abnormally high proton conductance of the pore. If in fact proton movement through the pore proceeds by its hopping from one water molecule to another down a chain, little coupling of proton movement to water flow is expected. The absence of measureable streaming potentials confirms this expectation.

Two determinations of N (the number of water molecules in the gramicidin A pore) have been made, based on two different, and totally independent, experiments and theories. The first [eq.(8-2)] results from a comparison of P_f and P_{d_w}, and derives from a theory of water-water interactions in a single-file pore. The second [eq.(8-3)] results from streaming potential measurements and comes from considerations of single-file interactions of water and ions. Given the experimental difficulties attendant to both determinations, the agreement between the values in eqs.(8-2) and (8-3) is quite satisfactory and can even be viewed as independently confirming the single-file (or no-pass) aspect of ion and water transport through the gramicidin A pore; the value of 5.3 for N comes from the assumption that water molecules cannot pass each other within the pore, whereas the value of 6.5 derives from the assumption that water molecules cannot pass ions (and vice versa) within the pore.

The value of 6 [eqs.(8-2) and (8-3)] for the number of water molecules in the gramicidin A pore is a reasonable one. The β_6-helix model of the pore (Urry et al., 1975), a cylinder 2 Å in radius and approximately 25 Å in length, maximally accommodates only 10 water molecules (Levitt et al., 1978). Furthermore, recent molecular dynamics calculations (Mackay et al., 1984) and Monte Carlo simulations (Fornili et al., 1984) of ion and water behavior in this pore model indicate a value of 8 to 9 water molecules in the ion-free pore and 7 to 8 in a pore with single-ion occupancy.

The details of the water structure within the pore that emerge from these analyses are very interesting. The water molecules are hydrogen-bonded together in an array with parallel dipoles oriented in the same direction, even in the ion-free pore; movements of the water molecules are highly correlated throughout the length of the pore, particularly in association with ion movement. Although at the molecular level the structure and behavior of water within this single-file pore appears very different from that of bulk water, it should be recalled that the transport rate of water through the pore is in reasonable accord with macroscopic theory. We previously noted the concordance of the experimental values of p_f with that predicted from Poiseuille's law. It is also noteworthy that the replacement of H_2O by D_2O reduces single-channel conductances by about the same amount (15–20%) that it reduces ionic mobilities in bulk solution (Tredgold and Jones, 1979). The latter effect is explained simply by the greater viscosity of D_2O. It therefore appears that despite the subtleties of hydrogen bonding and dipole orientations, the effect of D_2O on single-channel conductance can be described in terms of its effect on the viscosity of the solution within the pore.

Effect of Ions on the Water Permeability of the Pore

The values of P_f and P_{d_w} considered in an earlier section were those of an ion-free gramicidin A pore. That is, the water permeability experiments were performed at salt concentrations for which the probability of ion-occupancy of a pore was small; in other words, under the experimental conditions, most of the pores at any instant in time were ion-free. Ordinarily, the presence of ions in a pore would be expected to introduce only second-order alterations in water permeability coefficients, but because the no-pass condition in a single-file pore intimately relates ion and water movement, profound effects of ion occupancy on water permeability can occur. Conversely, the water permeability places constraints on ion transport and channel conductance, because of the no-pass condition. In this section, we first examine the predictions made from single-file theory concerning the effects of ions on water permeability and of water permeability on ion transport, and then compare them with the experimental results obtained on the gramicidin A pore.

Theory. Before beginning our discussion of the effect of ion occupancy on water permeability, a technical point with respect to the experimental determination of P_f must be considered. In principle, the value obtained for P_f of a porous membrane depends on the experimental conditions of the measurement. In particular, under open-circuited conditions, the streaming potential has a retarding effect on osmotic water flow (known as the electroviscous effect and actually resulting from a back electroosmotic component opposing the pressure-generated flow). Hence, P_f measured under open-circuited conditions will be less than that measured with the potential shorted out under closed-circuited conditions [see Lorenz (1952) for a discussion of this issue]. Usually, the effect of streaming potentials on water flow is negligible, and consequently it is generally not necessary to specify the conditions under which P_f is measured. Streaming potentials have a profound effect, however, on water flow through a pore in which single-file transport occurs. Indeed, for a membrane containing cation (or anion) permselective single-file pores, no osmotic water flow can occur in the open-circuited condition through ion-occupied pores; if it did, charge would be continually transferred across the membrane from one compartment to the other, thereby producing macroscopic violations of electroneutrality* [see Rosenberg and Finkelstein (1968a) for a more complete analysis of this situation].

The hydraulic permeability coefficient, L_p [eq.(2-1b)], and hence P_f [eq.(2-3)], relate to the volume flow occurring under short-circuited conditions [see, for example, de Groot (1958)]. We can see from the above considerations that this makes sense, as P_f is supposed to be a measure of the volume flow resulting from a hydrostatic (or osmotic) pressure difference *alone,* uncomplicated by electro-osmotic flow induced by electrostatic potential differences. Certainly, this is the only meaningful definition of P_f that permits its comparison to P_{d_w}. In the following discussion, therefore, of single-file pores in general and the gramicidin A pore in particular, it should be understood that references to P_f and its values always pertain to the short-circuited condition. The above issue does not arise with P_{d_w}, since it generally is determined in symmetric salt solutions with no potential difference across the membrane; hence, the "short-circuited" condition automatically applies to P_{d_w} measurements. Most of the discussion below, in fact, will be in reference to P_{d_w} rather than to P_f.

---

*This statement is strictly true if all ion-occupied pores contain the same number of ions. If, for example, some contain one ion and others contain two, an opportunity exists for local currents to flow among the pores (if the streaming potentials for the singly occupied and doubly occupied pores are different), and hence net osmotic flow of water can take place in the open-circuited condition.

The movement of an individual water molecule through an ion-free, single-file pore necessitates the movement of all N water molecules through the pore; its movement through an ion-occupied pore requires, in addition to the movement of the N water molecules,* transport of the ion through the pore as well. If the added resistance to movement contributed by the ion is small compared to that contributed by the N water molecules, the values of the water permeability coefficients for the ion-occupied and ion-free pore will be the same. On the other hand, if the added ion resistance is comparable to or larger than that of the N water molecules, the values of P_{d_w} and P_f for the ion-occupied pore will be lower than those for the ion-free pore. In the limit, the ion plugs the pore, and P_{d_w} and P_f become zero—the pore is impermeable both to ions and to water.

We have just considered the effect of ion permeability on water permeability, but the reverse effect is equally interesting. If the resistance contributed to ion movement by electrostatic barriers and ion-wall interactions is large compared to the frictional resistance associated with the movement of N water molecules through the pore, no particular relationship exists among the g_{max}'s for different ions. [g_{max} is the maximum (small signal) single-channel conductance, achieved at high salt concentrations when the pore always contains one ion.†] On the other hand, if it is small in comparison to the resistance contributed by the N water molecules, the g_{max}'s for all ions are the same. In fact, the rate at which an ion once in the pore moves through it is not only identical for all ions, it is also the same as that of a tagged water molecule in an ion-free pore, since in both cases the major resistance to transport experienced by the species comes from the necessity to move N water molecules along with it through the pore.

These qualitative statements can be made quantitative. In particular, it can be shown (Finkelstein and Andersen, 1981) that the

---

*For simplicity, we assume there are the same number of water molecules (N) in an ion-occupied pore and ion-free pore.

†For simplicity, we do not consider multiple-ion occupancy.

maximum possible unidirectional flux* of an ion, $(M_{ion})^{m.p.}$, (occurring at high salt concentrations when the pore always contains an ion) is related to the unidirectional flux of water through an ion-free pore, $(M_w)_{[ion]=0}$, by the equation:

$$(M_{ion})^{m.p.} = \frac{1}{N} (M_w)_{[ion]=0} \tag{8-4}$$

This is the maximum possible unidirectional flux of an ion, because the only impediment to ion movement assumed in its derivation comes from the necessity for N water molecules to move through the pore with the ion. Additional relations between ion conductance and water permeability can be derived [see, for example, Finkelstein and Andersen (1981); Dani and Levitt (1981b); Hladky and Haydon (1984)], but this theoretical aspect of the subject will not be pursued further here.

Gramicidin A Pore. The predicted inhibitory effect of ion occupancy on osmotically induced water flow in the open-circuited condition is experimentally observed with Li^+, K^+, and Tl^+, and was in fact used to determine binding constants of these ions in the gramicidin A pore (Dani and Levitt, 1981a). These observations further confirm the single-file nature of ion and water transport in this pore. In contrast, although Na^+ occupancy should produce about a four-fold reduction in P_{d_w} [see eq.(8-9) below and combine it with eq.(8-4)], no effect of Na^+ concentration on P_{d_w} was observed experimentally by Finkelstein (see Finkelstein and Andersen, 1981). Unfortunately, the osmotic studies of Dani and Levitt (1981a) did not include Na^+, and the tracer flux experiments of Finkelstein did not include any of the ions tested by them, so the exact status of these

---

*The unidirectional flux of a species is the flux that would be measured if all of its molecules on one side of the membrane were isotopically labeled and none were labeled on the other side. In practice, only a small fraction are labeled on one side, and the unidirectional flux is calculated by dividing the measured flux by the specific activity of the isotopic label.

findings is not certain. One possibility raised by Finkelstein and Andersen (1981) is that water molecules can pass a sodium ion somewhere in the pore; i.e., the no-pass condition is not fulfilled in all parts of the pore. They suggested that strict single-file transport of ions and water occurs between the two energy wells at the ends of the pore (see Fig. 8-3), but water molecules can pass a sodium ion sitting in either well. The most likely place for these wells is at the first turn of the β_6-helix, and they showed that it is possible for a sodium ion to bind in that region and still permit water molecules to pass. Presumably, the binding sites for the other alkali cations are slightly further into the pore, where they block the passage of water molecules around them.

One of the most interesting aspects of the study of water permeability in the gramicidin A pore is the light it sheds on the energy barriers to ion transport (Fig. 8-3). Inserting the value of 1.82×10^{-15} cm^3/s for p_{d_w} [obtained in 0.01 M NaCl on PE membranes (Rosenberg and Finkelstein, 1978b)] into the relation:

$$M_w = p_{d_w} c_w \qquad (8\text{-}5)$$

where c_w is the concentration of water in water ($= \frac{1}{18}$ mol cm^{-3}), we have for the unidirectional flux of water in an ion-free pore (in 0.01 M NaCl a pore is almost always ion-free):

$$(M_w)_{[\text{Na}^+]=0} = 6.1 \times 10^7 \text{ molecules/s} \qquad (8\text{-}6)$$

Substituting this value into eq.(8-4) with $N = 6$ gives for the maximum *possible* unidirectional flux of sodium:

$$(M_{\text{Na}})^{\text{m.p.}} = 1.0 \times 10^7 \text{ ions/s} \qquad (8\text{-}7)$$

On the other hand, from the maximum (small signal) single-channel sodium conductance ($g_{\max}$) of 14.6 pmho obtained for the gramicidin A pore in PE membranes at high NaCl concentrations (Finkelstein and Andersen, 1981), the unidirectional flux of sodium through a pore that always contains one sodium ion (in other words, the maximally *observed* unidirectional flux of sodium) is calculated to be (see Finkelstein and Andersen, 1981):

$$(M_{Na})_{[Na^+]=\infty} = 2.3 \times 10^6 \text{ ions/s} \qquad (8\text{-}8)$$

Comparing eq.(8-8) with eq.(8-7) we have:

$$\frac{(M_{Na})^{m.p.}}{(M_{Na})_{[Na^+]=\infty}} = 4.3 \qquad (8\text{-}9)$$

Equation (8-9) is rather remarkable. (It states that the unidirectional sodium flux through a pore that is always occupied by a sodium ion, $(M_{Na})_{[Na^+]=\infty}$, is only about four-fold smaller than the maximum possible unidirectional sodium flux, $(M_{Na})^{m.p.}$, calculated from the water permeability of an ion-free pore. This means that Na^+ flux through the pore is limited primarily by the frictional resistance experienced by the six water molecules that are constrained to move with the ion because of the single-file nature of transport. In fact, the rate of Na^+ movement between the wells of Figure 8-3, calculated by Andersen and Procopio (1980) from independent electrical measurements, is in complete agreement with eq.(8-7). Thus, there is no significant electrostatic energy barrier between the wells (Fig. 8-3B), and the frictional resistance of Na^+ with the walls of the pore is small compared to that of the six water molecules that must move with it; in other words, its rate of movement between the two wells is totally determined by the rate of moving six water molecules through the pore. The maximally *observed* sodium flux falls short of the maximum *possible* sodium flux [see eq.(8-9)] only because of the limitation imposed by the exit rate constant for Na^+ from the wells at the ends of the pore into the adjacent solutions [see Fig. 8-3B and Finkelstein and Andersen (1981)].

Dani and Levitt (1981b) reach essentially the same conclusion based on somewhat different theoretical calculations and experimental data. From electrostatic considerations, Levitt (1978a) calculates a small electrostatic energy barrier of about $4\,kT$ between the wells. Using this number, their value of $6 \times 10^{-14}\,\mathrm{cm^3/s}$ for p_f (Dani and Levitt, 1981a), and $N = 9$, Dani and Levitt (1981b) also conclude that the diffusion coefficient of Na^+, K^+, and Tl^+ within the pore is the same as that for the row of N water molecules that must move with the ion, and that local interactions of ions with the walls are so small that they do not significantly affect channel conductance. [Li^+ is an apparent exception to this statement; its conductance is probably limited by local ion-wall interactions (Dani and Levitt, 1981b).] The different g_{max} values for these ions result from differences in their exit rate constants from the wells at the ends of the pore. Additional comments on ion-water interactions within the gramicidin A pore can be found in the review by Hladky and Haydon (1984).

Dani and Levitt (1981b) also conclude that the intrinsic diffusion coefficient of water within the pore is about equal to the self-diffusion coefficient of water in bulk solution. The intrinsic diffusion coefficient is the hypothetical diffusion coefficient that one water molecule would have if it were alone in the pore (Levitt and Subramanian, 1974); it reflects intrinsic water-pore wall interactions in the absence of water-water interactions. Thus, water-pore wall interactions are comparable to water-water interactions in bulk water. This may be another aspect of why Poiseuille's law works so well in calculating p_f for this narrow pore.

E. COMPARISON OF THE GRAMICIDIN A PORE WITH THE NYSTATIN AND AMPHOTERICIN B PORE

It is instructive to compare the water permeability of the 2 Å-radius gramicidin A pore with that of the 4 Å-radius nystatin and amphotericin B pore, particularly since these are the only pores in lipid bilayer membranes, or plasma membranes, for which more or less complete information is available. Table 8-1 summarizes the water permeability data along with additional relevant information pertaining to these pores. Most of the points that can be made from the entries in the table have already been made in this and the preceeding chapter, but it is useful to recapitulate three of them here in one section. First, although the radius of the gramicidin A pore is a factor of 2 smaller than that of the nystatin and amphotericin B pore, P_f/P_{d_w} is larger for the gramicidin A pore. Although this is contrary to the general trend of P_f/P_{d_w} declining with decreasing pore radius, it is consistent with the single-file nature of transport through the smaller pore. Second, the values of p_f and p_{d_w} for these pores of molecular dimension are within an order of magnitude of the values predicted from macroscopic theory. Third, although this monograph does not focus on ion permeation, it is worth noting that the conductance of the 2 Å-radius gramicidin A pore is almost two orders of magnitude greater than those of the 4 Å-radius nystatin and amphotericin B pores. This illustrates the danger of trying to infer, as is often done, channel radius from channel conductance. Electrostatic forces can obviously play a dominant role in determining channel conductance if there are charges associated with the pore walls. They are also of critical importance, however, even if there are no charge groups associated with the pore walls, as in the present two examples.*

---

*The carboxyl and amino groups that sit at the ends of the polyene pores apparently do not significantly affect ion permeability (Dennis et al., 1970; Finkelstein and Holz, 1973).

TABLE 8-1. Comparison of Gramicidin A Pore with Nystatin and Amphotericin B Pores

Pore	Radius (Å)	Length (Å)	Conductance in 100 mM KCl (mho)	P_f/P_{d_w}	p_f(cm³/s) Experimental	p_f(cm³/s) Poiseuille's Law
Gramicidin A	2	25–30	10^{-11}[a]	5.3[d]	$1\text{-}6 \times 10^{-14}$[d,e]	3×10^{-14}
Nystatin (single-length)	4	21–25	2.5×10^{-13}[b]	—	10×10^{-14}[f,g]	62×10^{-14}
Nystatin (double-length)	4	42–50	1.3×10^{-13}[b]	3.3[g]	1.5×10^{-14}[f,g]	31×10^{-14}
Amphotericin B (double-length)	4	42–50	1×10^{-12}[c]	3.0[g]	4.5×10^{-14}[c,g]	31×10^{-14}

[a]Hladky and Haydon (1972).
[b]Kleinberg and Finkelstein (unpublished observations).
[c]Ermishkin et al. (1977).
[d]Rosenberg and Finkelstein (1978b).
[e]Dani and Levitt (1981a).
[f]Kleinberg and Finkelstein (1984).
[g]Holz and Finkelstein (1970).

PART
III

PLASMA MEMBRANES

This part of the monograph is divided into three chapters. The first, Chapter 9, considers the question of how water crosses the plasma membrane of cells *in general*; is it by a solubility-diffusion mechanism through the bilayer proper of the membrane, or is it through aqueous pores? Drawing from the theoretical concepts in Part I and from the results on lipid bilayer membranes in Part II, I present and discuss several criteria that can be used to decide this question. In the last two chapters, these considerations are applied to two specific systems: the red cell membrane (Chapter 10) and epithelial cell membranes (Chapter 11), with particular reference in the latter chapter to the water permeability pathway induced in toad urinary bladder and cortical collecting tubules of the mammalian kidney by the antidiuretic hormone (ADH).

9

General Considerations

Because of the presumed insolubility of water in oil and the relatively large water permeability of cell membranes in comparison to their polar nonelectrolyte permeability, it is usually assumed by the laity that water must cross cell membranes through pores. As discussed in Chapter 6, however, water has a finite solubility in hydrocarbon, and the water permeability of lipid bilayer membranes is adequately explained by the hydrocarbon-like properties of the bilayer interior. Moreover, since the values of the water permeability coefficients, which are dependent on the lipid composition of the bilayer, span a range that includes most of those reported for cell membranes, there is no difficulty in accounting for the large magnitudes of the water permeability coefficients of plasma membranes simply in terms of the properties of their lipid bilayer matrices. [It is in fact amusing that the values most out of line with those obtained on lipid bilayers are at the low end of the spectrum. For instance, the value of 1×10^{-6} cm/s for P_{d_w} of the plasma membrane of *Fundulus* eggs (Dunham et al., 1970) is 20-fold lower than the smallest value obtained on planar lipid bilayer membranes, and presumably results from a lipid bilayer composition that is less fluid (perhaps by being below the phase transition) than any thus-far studied in planar bilayers.] The existence of ion-conducting channels and other transport systems in plasma membranes, however, provides additional potential pathways for water movement that cannot be dismissed *a priori*. Indeed, it is almost certain that water is permeant through most of these regions, and therefore the question is a quantitative one: which is the *predominant* route that water utilizes in crossing plasma membranes? Is it through the lipid bilayer, or through polar transport sites (pores)? Related to this question is the problem of establishing criteria for distinguishing experimentally between these two possibilities. These are the issues addressed in this chapter.

A. THE RATIO OF P_f TO P_{d_w}

The least ambiguous criterion *in principle* for deciding whether osmotic water transport* occurs primarily by a solubility-diffusion process through the lipid bilayer of the plasma membrane or by quasi-laminar flow through a polar, porous pathway is the value of P_f/P_{d_w}. As was extensively discussed in Parts I and II, $P_f/P_{d_w} = 1$ if the route of water movement is through an oil membrane (e.g., a lipid bilayer), whereas $P_f/P_{d_w} > 1$ if water traverses a membrane through pores. [An exception to this statement is transport through a single-file pore containing only one water molecule; for that situation, $P_f/P_{d_w} = 1$; see eq.(4-17).]

The first measurements of P_f/P_{d_w} made on biological (or any) membranes were by Hevesy et al. (1935) on frog skin (attached to the whole frog!) shortly after methods were developed for obtaining isotopically enriched water. They discovered, much to their surprise, that its value was greater than 1. The possibility of using the magnitude of this ratio both to establish the presence of pores in membranes and to characterize their size was recognized and introduced into the biological literature about 20 years later in the classic papers of Koefoed-Johnsen and Ussing (1953) and Pappen-heimer (1953); it was subsequently taken up with enthusiasm by a number of investigators and applied by them to various cell types

---

*The water transport across the plasma membrane that is of relevance to the cell is *net* water movement, that is, osmotic transport. Tracer exchange, although revealing to the physiologist and biophysicist, is of no physiological consequence to the cell. For this reason, I have phrased the issue in terms of osmotic water transport. From the physical standpoint, it could be equally well expressed in terms of water movement *per se,* without specific reference to either osmotic or tracer flux.

(e.g., Prescott and Zeuthen, 1953; Paganelli and Solomon, 1957; Villegas and Villegas, 1960). For some cells such as *Xenopus* egg and *Amoeba chaos chaos,* the value of P_f/P_{d_w} was found to be close to 1 (Prescott and Zeuthen, 1953), but for many others it was clearly greater than 1, and in some, such as frog and zebra fish ovarian eggs (Prescott and Zeuthen, 1953), it was orders of magnitude greater, thus implying the presence of very large pores in those membranes.

The validity of this entire approach to determining the existence and size of pores in membranes, however, was severely challenged by Dainty (1963), who argued that the presence of unstirred layers could result in serious underestimations of P_{d_w}, thereby generating spuriously high values of P_f/P_{d_w} (see Chapter 3). [Ironically, the issue of unstirred layers was raised in the original Hevesy et al. (1935) paper.] This valid critique of P_f/P_{d_w} determinations places on the investigator claiming to measure a value significantly greater than 1 the onus of proof that his finding is not an artifact of unstirred layers; any reports of $P_f/P_{d_w} > 1$ that do not adequately address this point must be regarded with suspicion. As a rule of thumb, findings of P_f equal to P_{d_w} [e.g., *Valonia* (Gutknecht, 1967)] are *a priori* believable, whereas claims of P_f greater than P_{d_w} [e.g., squid axon (Nevis, 1958)] are *a priori* suspect. (Chapters 10 and 11 each present an example of a plasma membrane for which the finding of $P_f > P_{d_w}$ is not an artifact of unstirred layers.*) This state of affairs is generally accepted and recognized by the *cognoscenti,* and its truth limits the utility of the P_f/P_{d_w} criterion for determining the primary route of water transport across cell membranes. Consequently, other criteria, discussed in the sections below, have been proposed for answering this question. Since the measured value of P_{d_w} may not accurately reflect the true value of the diffusional water permeability coefficient for the cell membrane, because it is confounded by the (negative) contribution of unstirred layers to its magnitude, attention is focused in these sections on P_f in analyzing and discussing water permeability data.

---

*The issue of an unstirred layer artifact reemerges in Chapter 11 in a more subtle form.

B. TEMPERATURE DEPENDENCE OF P_f

The effect of temperature on water permeability can, with caution, he used to identify the nature of the water-permeation pathway. For the solubility-diffusion mechanism of transport through a lipid bilayer, the Q_{10} for P_f is about 2 (see Chapter 6). In contrast, for water flow through an aqueous pore, one expects, in conformity with macroscopic theory, that its Q_{10} will essentially be that for the viscosity of water, which is about 1.25.* There has been, unfortunately, no explicit determination of the temperature dependence of P_f for the polyene and gramicidin A pores[†], the only pores for which water permeability coefficients have been measured. The Q_{10} for the single-channel conductances of the gramicidin A pore, however, has been measured, and is about 1.35 (Hladky and Haydon, 1972). This implies, given the intimate association of ion and water transport through the gramicidin A pore discussed in Chapter 9, that the Q_{10} for P_f is similarly small. It would therefore appear that a clear distinction between water transport through the bilayer proper and water transport through pores could be made on the basis of the temperature dependence of P_f.

The major drawback to this approach is illustrated by the polyene and gramicidin A pores themselves. The *number* of these pores in the membrane and their mean open-time duration is significantly affected by temperature (Cass et al., 1970; Hladky and Haydon, 1972); consequently the Q_{10} for the macroscopically measured osmotic permeability coefficient of the membrane (P_f) does not reflect the intrinsic Q_{10} for the osmotic permeability coefficient of the pore itself (p_f). Of course, for a membrane containing these pores, whose Q_{10}'s for single-channel conductances are known (or can be determined), it is possible to normalize P_f values for the number of open pores

---

*If P_{d_w} is unstirred-layer limited, its Q_{10} will have this value regardless of the mechanism of water transport through the membrane, since what is being measured is simply the temperature dependence of the self-diffusion coefficient of water in water.
[†]A low Q_{10} was determined for P_f of gramicidin A-treated liposomes (Boehler et al., 1978), but there is no way of knowing whether the number of pores or their mean open time remained constant with temperature (see the remarks in the following paragraph).

in the membrane and thereby obtain the intrinsic temperature dependence of p_f. For the plasma membrane, however, which contains a variety of pores and transport systems and where the question of water traversing the membrane through pores is at issue in the first place, one does not have this luxury. Thus, the possibility of temperature affecting pore number and lifetime, as well as having other potential actions on plasma membrane structure and composition, complicates the interpretation of temperature effects on water permeability, and hence usually does not permit an unambiguous conclusion based on Q_{10} measurements to be drawn concerning the water permeability pathway.

C. VALUES OF REFLECTION COEFFICIENTS (σ's)

If the reflection coefficient (σ) for a solute s is less than 1 [or, actually, less than $(1-\omega \bar{V}_s/L_p)$, where $\omega \equiv P_{d_s} A/RT$ (eq.(5-8)) and $L_p \equiv P_f A \bar{V}_w/RT$ (eq.(2-3))], it indicates that solute and water molecules interact as they cross the membrane, which in turn implies that both permeate the membrane through aqueous pores [see eq.(5-10) and the analysis of σ in Chapter 5]. Thus the finding for a solute, or solutes, that

$$\sigma < 1 - \frac{\omega \bar{V}_s}{L_p} \tag{9-1}$$

is evidence that the major route of water transport across the membrane is through pores.*

There are a number of potential problems that restrict the usefulness of this inequality as a criterion for water transport through aqueous pores. First, if the pores are so narrow that even small solutes are ex-

---

*Although the meaning of $\sigma = (1-\omega \bar{V}_s/L_p)$ is that solute and water do not interact or "see" each other as they cross the membrane, it does not necessarily mean that water traverses the membrane by a solubility-diffusion mechanism through the bilayer. For example, the equality will also be satisfied if the solute permeates the membrane by a solubility-diffusion mechanism or by a special carrier system, while water traverses the membrane through aqueous pores.

cluded, as in the case of the gramicidin A pore, no solutes will be found for which the inequality applies. Second, unless the pores are very wide, σ values will tend to be close to 1, making it experimentally difficult to establish whether the inequality in (9-1) is satisfied or not. Third, there are several possible ways of obtaining artifactually small values of σ (Levitt, 1974; Levitt and Mlekoday, 1983), which then leads to the erroneous conclusion that inequality (9-1) holds, when in fact it does not. Finally, if the test solute is moderately permeant (through the bilayer or through an independent transport system), unstirred layers can reduce its concentration difference across the membrane proper, thereby reducing the osmotic flow rate produced by it; this can be incorrectly interpreted as resulting from a low value for σ. These problems limit the usefulness of reflection coefficient measurements as a tool for establishing the pathway of water transport across a cell membrane.

D. SOLVENT DRAG

Another manifestation of solvent-solute interaction in a porous membrane is the solvent drag effect. If a solute is permeant through aqueous pores in a membrane, then in the face of water flow through those pores (induced by either a hydrostatic pressure difference or an osmotic pressure difference of an impermeant solute across the membrane), solute flux will be increased in the direction of water flow and will be retarded in the opposite direction; the solute is "dragged" through the pores by the solvent. No such coupling of solvent and solute fluxes occurs with the solubility-diffusion mechanism of their transport through a lipid bilayer. This feature of solvent-solute coupling was discussed in the last section of Chapter 5 in conjunction with the measurement of σ by ultrafiltration experiments, and in fact the solvent drag effect is another manifestation of σ values that are less than $1 - \omega \bar{V}_s / L_p$ [inequality (9-1)].

The major experimental difficulty in demonstrating a true solvent drag effect, and hence in establishing the existence of a porous pathway for water flow across a membrane, is that the presence of unstirred layers can give rise to a "pseudo" solvent drag; that is, a concentration gradient of solute is created across the membrane proper in the direction of water flow, thereby generating an increased flux of solute through the bilayer proper, or through an independent transport pathway, that mimics solvent drag. This phenomenon is extensively analyzed by Barry and Diamond (1984), and the interested reader should consult their paper for details. Basically, the more permeant the solute and the smaller its σ, the larger is the effect of unstirred layers on its flux, and hence the greater is pseudo solvent drag. On the other hand, for poorly permeant solutes with σ's close to 1, the true solvent drag effect is small and difficult to measure.

Experimental studies of solvent drag have been carried out for the most part on epithelia. Barry and Diamond (1984) review the literature and conclude that in none of the purported claims of solvent drag effects can pseudo solvent drag be precluded as the source of the observations. Andreoli et al. (1971) attempted to measure solvent drag on amphotericin B-treated planar bilayer membranes. It is instructive to observe the contortions these authors go through with the unstirred layer problem in order to try to demonstrate solvent drag in this well-defined system in which one not only knows *a priori* that solvent drag must occur, but also knows the magnitude of the expected effect. They ultimately claim to observe the predicted solvent drag effect, but I find their experiments and analysis unconvincing.

E. COMPARISON OF P_f TO SOLUTE P_d's

We saw in Table 6-1 of Chapter 6 that the relative P_d values of solutes and water in lipid bilayer membranes are essentially constant, even though their absolute values change by orders of magnitude with lipid composition and temperature. In particular, for any solute s tested, the value of P_{d_w}/P_{d_s} (and hence P_f/P_{d_s}) was more or less independent of bilayer composition. This finding prompted me (Finkelstein, 1978b) to propose this as a general rule for all bilayers, including those of plasma membranes. Thus, if water traverses a plasma membrane by a solubility-diffusion mechanism through the lipid bilayer, P_f/P_{d_s} should approximate the value found in artificial lipid bilayer membranes (Table 6-1), where s is a solute that permeates the plasma membrane without aid of special transport systems; that is, s is a solute that also traverses the membrane by a solubility-diffusion mechanism through its bilayer. (Examples of likely appropriate test solutes that meet this condition are 1,6-hexanediol, 1,4-butanediol, n-butyramide, and isobutyramide.) Conversely, if the value of P_f/P_{d_s} greatly exceeds the corresponding value in Table 6-1, it is a strong indication that the primary route of water transport is through aqueous pores. The test solutes, of course, should not utilize these pores as a major pathway; this could be established by confirming that the ratios of their P_d's approximate the corresponding ratios in Table 6-1.

F. THE NUMBER OF PORES IN A MEMBRANE

With the exception of epithelial cells, most cells do not need or require, from a physiological standpoint, high water permeabilities, and whatever needs they do have can readily be met by the intrinsic water permeability of the lipid bilayers of their plasma membranes. One therefore does not expect (again, with the exception of epithelial cells) that the plasma membranes of most cells have evolved special pores for water transport. Consequently, it can be argued that if a large fraction of the water movement across a particular plasma membrane occurs through pores, it must do so incidental to other transport properties of those pores (for instance, ion movement) and is not representative of their primary function. If, therefore, an estimate can be made of the number of pores in a cell membrane devoted to specific transport functions, one can ask if this is a sufficient number to significantly contribute to water movement across the membrane. Because of the large variety of cells and pore types, it is not reasonable to expect a single, all embracing answer to this question, but a useful, general outlook is provided by approaching the question from this viewpoint.

As an illustration of this approach, let us consider a plasma membrane having the following properties:

$P_f = 2 \times 10^{-3}$ cm/s   (osmotic permeability coefficient) (9-2a)

$G = 10^{-3}$ mho/cm^2 (membrane conductance)   (9-2b)

$g = 10^{-11}$ mho   (single-channel conductances)   (9-2c)

$n = \dfrac{G}{g} = 10^8/\text{cm}^2 = 1/\mu\text{m}^2$ (density of channels in membrane)   (9-2d)

The values in eq.(9-2) reflect my subjective impression of "typical" values reported in the literature. [The plasma membrane of the resting squid giant axon—with $G = 4 \times 10^{-4}$ mho/cm^2 (Hodgkin et al., 1952), $g_{K^+} = 1.8 \times 10^{-11}$ mho (Conti and Neher, 1980), $g_{Na^+} = 4 \times 10^{-12}$ mho (Conti et al., 1975), and $P_f = 1.1 \times 10^{-3}$ cm/s (Villegas and Villegas, 1960)—approximates this canonical membrane.] Are there enough pores in this membrane to account for its water permeability? To answer this question some estimate has to be made of the water permeability *per pore,* and to make this estimate we draw on the only information presently available—namely, the water permeability of the polyene and gramicidin A pores. The reason one presumes it possible to extrapolate from the water permeabilities of these pores to those of biological channels is that since the p_f's of the former are in reasonable agreement with the predictions from macroscopic hydrodynamic theory, there is some ground for believing that the same considerations are applicable to any pore of comparable dimensions.

We shall assume that the water permeability of plasma membrane channels is not significantly greater than (and probably in general is less than) that of the single-length nystatin pore. The rationale for this assumption is the following: The ion-selective regions of plasma membrane channels are probably considerably narrower than the 4 Å radius of the nystatin pore (see Hille, 1975), but they may be somewhat shorter than its 25 Å length (see Miller, 1982). The shorter length would tend to make the p_f of ion-selective channels larger than that of the nystatin pore, but this should be more than compensated for by their narrower width. Taking, therefore, $p_f = 10^{-13}$ cm^3/s (see Table 7-2) as an upper limit for the water permeability per channel and multiplying this by n, the number of channels in the membrane [eq.(9-2d)], we find that the value of P_f predicted from water transport through the channels is 10^{-5} cm/s, or less than 1% of the actual value for the plasma membrane [eq.(9-2a)]. Putting it another way, for pores to account for the water permeability of our "typical" plasma membrane, the membrane should have approximately 2×10^{10} channels/cm^2 ($200/\mu$m^2), whereas in fact it possesses only about 10^8 channels/cm^2 ($1/\mu$m^2). In short, there are too few ion-conducting channels in most plasma membranes to act as a significant pathway for water movement; by implication, most of a membrane's water permeability is attributable to its bilayer structure.

The above argument is applicable to any cell membrane for which P_f and single-channel conductance data exist. It is most useful in estimating the contribution of a specific channel type to P_f, but is less reliable in determining whether or not P_f is attributable to the lipid bilayer of the plasma membrane, since low conducting proteinaceous pathways may be present at a high enough density to contribute significantly to water permeability. For example, there are about 8,000 Na-K ATPase molecules per μm^2 in the chloride cell of the killifish's gill (Almers and Stirling, 1984)—enough to make a large contribution to P_f if water can traverse this transport system. It is interesting to note that since there are about 300 sodium channels/μm^2 in the squid giant axon (see Hille, 1984), the water permeability of its membrane should be substantially larger at the peak of the action potential than at the resting potential. This is even more true at the node of Ranvier, which has ten times the density of sodium channels as that of the squid giant axon (see Hille, 1984).

G. SUMMARY

From the considerations presented in the preceeding sections, it is apparent that it is not always a simple task to identify the major pathway for water transport across a cell membrane. Because of the difficulties associated with the successful application of any of the criteria discussed in this chapter, there are few cell membranes for which a clear-cut conclusion can be drawn. Two examples of membranes for which it is certain that the major water permeation route is through pores are presented in the following chapters. Even in those extensively investigated cases, we shall see how difficult it is to characterize the properties of those pores and to identify the proteins in the plasma membrane that form them.

10

The Red Cell Membrane

Throughout the history of cell membrane research, no cell has played a more important or prominent role than the red cell. From the seminal paper by Gorter and Grendel (1925) proposing a lipid bilayer structure for plasma membranes to recent work on membrane protein attachment to cytoskeletal elements (Branton et al., 1981), the red cell has been one of the foremost objects of research. Its easy accessibility in large, homogeneous quantities and the absence of internal membrane-bound structures (in enucleated mammalian erythrocytes) as sources of contamination for plasma membrane preparations commend it as the cell of choice for the investigation of a host of physiological and biochemical phenomena. In no area has the red cell membrane been more preeminent than in the field of transport processes, and among these, that of water transport (the subject of this chapter) has held a particular fascination for biophysicists. In fact, much of the general interest in water movement through narrow pores, both among biologists, and physicists, stems from the finding, initially made by A.K. Solomon and his colleagues in the 1950s and 60s, that water crosses the plasma membrane of (some) erythrocytes through pores.

This chapter reviews both the evidence for water movement through pores in the red cell membrane and the permeability characteristics of those pores. It also considers the perplexing question of which of the proteins in the membrane can possibly be involved in their formation. Most of the data and ideas are presented and analyzed in terms of the criteria for pores discussed in the preceeding chapter; in a sense, the present chapter is an application of those general criteria to the specifics of the red cell membrane. For the most part attention is confined to the human erythrocyte, as this is the cell that has been most intensively investigated; where appropriate, however, comparisons to the red cells of other species is made.

A. THE VALUE OF P_f/P_{d_w}

The concept of pores in the human red cell membrane initially arose from a comparison of the results of experiments on osmotic water entry into cells (Sidel and Solomon, 1957) with those on tritiated water exchange (Paganelli and Solomon, 1957). The authors of the former paper obtained a value of 1.26×10^{-2} cm/s for P_f, whereas those in the latter paper, having obtained a value of 5.3×10^{-3} cm/s for P_{d_w}, drew attention to the significance of the fact that $P_f/P_{d_w} > 1$; in this case, $P_f/P_{d_w} = 2.5$. They concluded from this finding that water crossed the membrane through pores, and then proceeded to calculate from eq.(3-6c) an equivalent pore radius of 3.5 Å. Over the years, the values of P_f and P_{d_w} have changed somewhat, presumably reflecting improvements in experimental technique and analysis rather than human erythrocyte evolution. The presently accepted value for P_f is about 2×10^{-2} cm/s (Rich et al., 1968; Terwilliger and Solomon, 1981; Mlekoday et al., 1983; Moura et al., 1984),* whereas that for P_{d_w} is about 3×10^{-3} cm/s (see Brahm, 1982). This raises the value of P_f/P_{d_w} for the human red cell membrane to 6.7, and, as we shall see shortly below, the value for the pores themselves is even greater. [Two different methods have been used to determine P_{d_w}. In one, the exchange of isotopically labeled water across the membrane is measured, as discussed in Chapter 3. The other utilizes an NMR technique, in which the exchange of water

---

*There have been reports of rectification of water flow across the red cell membrane, with the magnitude of P_f measured for inward flow about 40% large than that measured for outward flow (e.g., Farmer and Macey, 1970). Mlekoday et al. (1983), however, attribute the apparent rectification to a light scattering artifact, which when eliminated results in P_f values independent of the direction of water flow. Terwilliger and Solomon (1981) also find no rectification.

is measured either by proton resonance relaxation in the presence of low extracellular concentrations of Mn^{2+} or by ^{17}O relaxation in the absence of Mn^{2+} (see, for example, Fabry and Eisenstadt, 1975). Brahm (1982) reviews the results obtained by both methods, which range in value from 2×10^{-3} cm/s to 4×10^{-3} cm/s.]

We discussed in Chapter 9 that reported values of P_f/P_{d_w} greater than 1 are often erroneous, because unstirred layers can cause substantial underestimations of P_{d_w}. This does not appear to be an issue for the red cell measurements. Both theoretical calculations (Rice, 1980) and experimental measurements (Sha'afi et al., 1967) indicate that the extracellular unstirred layer thickness, under the mixing conditions of the tracer exchange experiments, is less than 6 μm, which introduces no appreciable error in the value of P_{d_w}. Furthermore, there is no extracellular unstirred layer in the NMR determinations of P_{d_w}. The intracellular unstirred layer is also inconsequential, since the diffusional distance to the membrane is less than 2 μm, and the diffusion constant of water in the concentrated hemoglobin solution of the cell is not seriously reduced from its value in pure water (Redwood et al., 1974); moreover, Brahm (1982) obtains similar values for P_{d_w} in both whole cells and ghosts. Finally, Blum and Forster (1970) and Brahm (1983a) find that the combined unstirred layer thickness on the two sides of the membrane is 2 μm or less. Thus, the nonunity value of P_f/P_{d_w} is *not* an artifact of unstirred layers.

The value 6.7 for P_f/P_{d_w} is that for the entire plasma membrane and includes in its measurement water movement occurring both through pores and through the lipid bilayer of the membrane. To obtain the value of P_f/P_{d_w} for the pores themselves, the contributions to P_f and P_{d_w} from the permeability of the lipid bilayer should be subtracted. That is,

$$\left(\frac{P_f}{P_{d_w}} \right)_{pores} = \frac{(P_f)_{measured} - (P)_{bilayer}}{(P_{d_w})_{measured} - (P)_{bilayer}} \qquad (10\text{-}1)$$

where,

$$(P)_{\text{bilayer}} \equiv (P_f)_{\text{bilayer}} = (P_{d_w})_{\text{bilayer}} \qquad (10\text{-}2)$$

In 1970, Macey and Farmer (1970) made the important discovery that the sulfhydryl reagents p-chloromercuribenzoate (PCMB) and p-chloromercuribenzene sulfonate (PCMBS), at concentrations of about 1 mM, reduce P_f by a factor of 10.* In subsequent experiments, Macey and his colleagues found that P_{d_w} was also reduced, but by less than a factor of 2, and, most interestingly, that at PCMBS concentrations at which water transport was maximally inhibited ($\sim$ 1mM), $P_f = P_{d_w}$ (Macey et al., 1972; Moura et al., 1984). Their interpretation of these findings was that PCMBS closes down the pores for water transport in the cell membrane, and that the remaining water permeability in PCMBS-treated erythrocytes is through the lipid bilayer proper of the membrane. This interpretation is supported by the change in the Q_{10} for P_f. In the untreated erythrocyte, the Q_{10} is about 1.2 (Vieira et al., 1970), consistent (as I remarked in the preceding chapter) with water flowing through aqueous pores.† In contrast, in the PCMBS-treated red cell, the Q_{10} of P_f is about 2 (Macey et al., 1972), which is the same as that for a lipid bilayer membrane (Graziani and Livne, 1972; Fettiplace, 1978).

In sum, it appears that about 90% of the osmotic flow of water across the human erythrocyte membrane occurs through pores and about 10% through the lipid bilayer; roughly equal amounts of tracer exchange occur through pores and bilayer. When the membrane is treated with 1 or 2 mM PCMBS, the porous route for water movement is completedly closed down, leaving only the bilayer

---

*The effects of these agents on solute and ion permeability are discussed in a later section.

†The Q_{10} for P_{d_w} is significantly larger (Vieira et al., 1970), but as Macey (1979) has pointed out, this can be attributed to the large Q_{10} of $(P_{d_w})_{\text{bilayer}}$ and to the fact that $(P_{d_w})_{\text{bilayer}}$ contributes much more to the overall P_{d_w} than does $(P_f)_{\text{bilayer}}$ to the overall P_f.

pathway.* One can therefore assume that the value of P_f and P_{d_w} obtained in PCMBS-treated erythrocytes is equal to $(P)_{\text{bilayer}}$ [eq.(10-2)], and use it in eq.(10-1) to calculate $(P_f/P_{d_w})_{\text{pores}}$, in which case one finds that:

$$\left(\frac{P_f}{P_{d_w}}\right)_{\text{pores}} \approx 10 \qquad \text{(human erythrocyte membrane)} \qquad (10\text{-}3)$$

(see Table 10-1). The experiments of Moura et al. (1984) support the position that, in terms of water permeability, PCMBS is simply closing down the pores (in an all or none manner), and not affecting the lipid bilayer. They show that the value calculated for $(P_f/P_{d_w})_{\text{pores}}$ remains constant at about 10 as the value of $(P_f/P_{d_w})_{\text{observed}}$ falls progressively from 5.9 to 2.1, when the PCMBS concentration is increased incrementally from 0.0 to 0.7 mM.

B. OTHER EVIDENCE FOR PORES

The large measured value of about 7 for P_f/P_{d_w}, even neglecting the correction of this number to 10 [eq.(10-3)] from the PCMBS experiments, is by itself convincing evidence that the major pathway for water flow across the red cell plasma membrane is through pores. The small Q_{10} for water permeation is further evidence for this, as we noted above. An additional finding consistent with a porous pathway is that P_f in D_2O is about 20% lower than in H_2O. As Karan and Macey (1980) remark, this difference is the same as that in the macroscopic viscosity of D_2O and H_2O solutions. (We noted in Chapter 8 that D_2O also produces about a 20% reduction in the conductance of the gramicidin A channel. Thus, once again, macroscopic theory appears to be applicable to pores of molecular

---

*Dix and Solomon (1984) take the position that virtually all water transport across the red cell plasma membrane, even in a PCMBS-treated cell, is protein mediated and that none occurs through the lipid bilayer. I feel that their arguments, which are indirect, are somewhat strained.

TABLE 10-1. *Water Permeability Coefficients for the Bilayer and Pore Pathways Across the Human Red Cell Membrane*

Pathway	Permeability Coefficients (cm/s $\times$ 10^{-2})	
Bilayer		
$(P)_{bilayer} \equiv (P_f)_{bilayer} \equiv (P_{d_w})_{bilayer}$	0.181	$(P_f/P_{d_w})_{bilayer} = 1$
Pore		
$(P_f)_{pore}$	1.97	$(P_f/P_{d_w})_{pore} = 10.6$
$(P_{d_w})_{pore}$	0.186	
Plasma Membrane		
$P_f [= (P_f)_{pore} + (P)_{bilayer}]$	2.15	$(P_f/P_d)_{membrane} = 5.9$
$P_{d_w} [= (P_{d_w})_{pore} + (P)_{bilayer}]$	0.367	

Source: From Moura et al. (1984).

dimension.) Significantly, in PCMBS-treated erythrocytes, essentially no difference is found between P_f measured in D_2O and P_f measured in H_2O (Karan and Macey, 1980), which is precisely what is expected, and found (Hanai and Haydon, 1966), for water flow through a lipid bilayer.

As a further manifestation of these pores' existence, we might expect the erythrocyte membrane to discriminate among small polar nonelectrolytes on the basis of size, in a manner similar to the "sieving" behavior of nystatin- and amphotericin B-treated membranes discussed in Chapter 7. In particular, this should be evidenced in nonunity reflection coefficient (σ) values [that is, σ values satisfying inequality (9-1)] for small solutes. This, in fact, is exactly what Goldstein and Solomon (1960) found. As we shall see in the next section, however, questions have been raised concerning this finding; moreover, there exist other observations difficult to reconcile with the proposition that water and small polar solutes share a common pathway.

C. NONELECTROLYTE PERMEABILITY

Although it has long been recognized that glucose crosses the human erythrocyte membrane through a specific transport system (see, for example, LeFevre, 1954), it appears at first glance that small polar nonelectrolytes, such as amides, ureas, and diols, show permeability patterns consistent with their being "sieved" through the water pores. This impression is obtained from both P_{d_s} (Sha'afi et al., 1971) and σ values (Goldstein and Solomon, 1960). The small σ value of approximately 0.6 for acetamide, urea, and ethylene glycol (Goldstein and Solomon, 1960) is particularly indicative of their traversing the membrane through the water pores. Upon closer scrutiny of solute permeability data, however, the argument for a shared permeation pathway for water and small polar solutes becomes shaky. Instead, the picture that emerges for the water pores is more like that for the gramicidin A pore—a pore so narrow (at least in some region) that only water, not even urea, is capable of passing through it.

Several lines of evidence have led to this conclusion. In the first place, the low σ values reported by Goldstein and Solomon (1960) are questionable. Levitt (1974) originally drew attention to a methodological error in their measurements that leads to underestimations of σ values. Subsequently, Owen and Eyring (1975) in a more refined series of measurements obtained σ's for urea, acetamide, and ethylene glycol of 0.79, 0.80, and 0.86, instead of the values of ≈ 0.6 reported by Goldstein and Solomon (1960). More recently, Levitt and Mlekoday (1983) in a painstaking experimental and theoretical treatment of σ data conclude that their results are best fit by $\sigma_{urea} = 0.95$ and $\sigma_{ethylene\ glycol} = 1.0$. They acknowledge that because of the complexity of their curve fitting procedures and calculations, their experiments do not provide accurate measurements of σ, and that σ_{urea} could be as low as 0.75 and still be consistent with their data. The important point of this study, however, which is the most careful thus far performed by anyone, is that there is *no* evidence from σ measurements that small solutes go through the water pores; the experimental results are perfectly consistent with urea and ethylene glycol using pathways separate from that of water.

A number of other observations also indicate that small solutes are not transported through the water pores. In their original paper, Macey and Farmer (1970) found that along with inhibiting water flow, PCMBS inhibits urea transport [which would be consistent with water and urea sharing a common pathway, except that the time and concentration dependence of these inhibitory effects are quite different (Levitt and Mlekoday, 1983)], but it does *not* inhibit ethylene glycol transport. Conversely, they found that phloretin inhibits urea transport by as much as a factor of 50 without significantly altering water (or ethylene glycol) permeability. Levitt and Mlekoday (1983) confirmed Macey and Farmer's (1970) results, and in addition discovered that copper inhibits ethylene glycol transport but not water (or urea) permeability. It is difficult to see how urea and ethylene glycol could be going through the water pores and yet there be ways of

(1) inhibiting water transport without affecting ethylene glycol permeability (PCMBS), (2) inhibiting urea transport without affecting water permeability (phloretin), or (3) inhibiting ethylene glycol transport without affecting water permeability (copper). The most obvious interpretation of these findings is that urea and ethylene glycol are not permeants of the water pores, but are transported instead via a separate pathway. [In fact, on the basis of saturation, competition, and inhibitor studies, Mayrand and Levitt (1983) conclude that urea and ethylene glycol do not share a common pathway, but rather permeate through separate transport systems.] The dissociation of water and urea transport is also supported by findings from comparative physiology. Thus, duck red cells have a high water permeability and a low urea permeability, whereas amphiuma red cells have a low water permeability and high urea permeability (Wieth and Brahm, 1977).

D. CHARACTERISTICS OF THE PORES

It appears inescapable from the evidence summarized in the preceeding paragraphs that the water pores of the red cell membrane transport water and little else. This is now the concensus among most workers in the field, and it has been lucidly articulated in a recent review by Macey (1984). [A contrary position is maintained by Solomon and his colleagues, who argue that water and small nonelectrolytes (and even cations and anions) go through the same pores, and that these pores have a radius of about 4.5 Å (Solomon et al., 1983). This viewpoint not only has the appeal of simplicity, it also deals most effectively with the problem of there possibly being too few proteins in the membrane to account for water, urea, ethylene glycol, and anion transport (see the following sections). Nevertheless, I find their Procrustian attempts to accommodate the evidence summarized above for separate solute and water pathways unconvincing, whereas Macey's (1984) detailed critique of their arguments is persuasive.] Thus, with respect to the large value of P_f/P_{d_w} and their impermeability to small nonelectrolytes, the water pores of the red cell membrane resemble those formed by gramicidin A, and water transport through them must occur by a single-file mechanism (at least in their narrowest region).

In one respect, however, the permeability properties of these pores differ significantly from those of the gramicidin A pore. As I originally pointed out (Finkelstein, 1974), the water pores of the red cell membrane are virtually ion impermeable, in contrast to the gramicidin A pore which is quite permeable to ions. This is most easily appreciated from macroscopic permeability and conductance measurements, without the necessity of invoking single-channel conductance data (of which there is none, and as we shall see, no prospect of obtaining any, for the red cell water pore). In 0.1 M NaCl at a gramicidin A-induced conductance of 10^{-2} mho/cm², the gramicidin A-induced P_f is 3.4×10^{-5} cm/s [Rosenberg and Finkelstein (1978b), assuming a linear extrapolation between $0.01 M$ and $0.1 M$ NaCl]. Therefore,

$$\frac{(P_d)_{\text{ion}}}{P_f} \approx 1 \qquad\qquad (10\text{-}4)^*$$

(gramicidin A-treated membrane in 0.1 M NaCl)

In contrast, the conductance of the red cell membrane in a similar salt solution is approximately 10^{-5} mho/cm² (Hunter, 1977), whereas $P_f \approx 2 \times 10^{-2}$ cm/s (e.g., Mlekoday et al., 1983), so that even if all of this conductance were attributable to the water pores:

*Calculated from the relationship between $(P_d)_{\text{ion}}$ and small signal conductance (G) at $V=0$:

$$(P_d)_{\text{ion}} = \frac{RT}{F^2} G \frac{1}{c}$$

where c is the salt concentration (Hodgkin and Keynes, 1955). This relation applies if the unidirectional ion fluxes satisfy the Behn-Ussing-Teorell flux-ratio equation (Teorell, 1953), which they do for Na^+ in gramicidin A-treated membranes (Procopio and Andersen, 1979).

$$\frac{(P_d)_{ion}}{P_f} \approx 10^{-6} \qquad \text{(red cell membrane in normal saline)} \qquad (10\text{-}5)$$

Comparing eqs.(10-5) and (10-4), we see that the water pores of the red cell membrane are at least 10^6-fold less permeable to ions than are the gramicidin A pores. Macey (1984) calculates a similar discrepancy between the H^+ permeability of the red cell water pores and that of the gramicidin A pores. The very low ion permeability of these pores is also characteristic of the antidiuretic hormone-induced pores in epithelia; in Chapter 11 we mention possible pore structures that combine a high water permeability with a low ion conductance.

E. WHO ARE THE WATER PORES?

In the preceeding chapter we made a *general* calculation of the number of pores required to account for the water permeability of a "typical" plasma membrane. Let us now make a similar *specific* calculation applicable to the human red cell membrane.

Given the single-file nature of water transport through the water pores, inferred from their exclusion of small nonelectrolytes, the finding of $P_f/P_{d_w} \approx 10$ [eq.(10-3)] implies that there are about 10 water molecules in single-file array in the narrowest portion of the pore. It is therefore reasonable to assume that the value of p_f for this pore is comparable to that of the gramicidin A pore; certainly, it should not be much greater. Setting p_f equal to 3×10^{-14} cm^3/s, the mean of the two determinations made on the gramicidin A pore (see Table 8-1), and $P_f \approx 2 \times 10^{-2}$ cm/s (e.g., Mlekoday et al., 1983), we have:

$$\frac{P_f}{p_f} \approx 7 \times 10^{11} \text{ pores/cm}^2 \qquad (10\text{-}6a)$$

(density of water pores in the human red cell membrane)

or given that the area of the red cell membrane is about $140 \, \mu m^2$ (Jay, 1975), this becomes:

$$n \approx 1 \times 10^6 \text{ pores/cell}$$
$$\text{(number of water pores in the human red cell membrane)} \quad \text{(10-6b)}$$

Solomon et al. (1983) perform a slightly different calculation, but arrive at a similar figure (2.7×10^5 pores/cell). There are thus roughly a million water pores in the human red cell membrane.

Which proteins are in sufficient copy number per cell to qualify as candidates for the water pore? (This can be uniquely asked of the red cell membrane; for no other plasma membrane is sufficient bio-chemical information available to intelligently pose the question.) On numerical grounds alone apparently only three proteins in the red cell membrane are eligible for consideration: band 3, band 4.5, and glycophorin,[*] of which there are about 1×10^6 (Knauf, 1979), 2.5×10^5 (Lin and Snyder, 1977), and 5×10^5 (Steck, 1974) copies per cell, respectively.[†] Glycophorin can probably be eliminated from consideration on two grounds: first, its intrabilayer domain consists of only a single strand of 22 amino acids (Marchesi, 1979), making it unlikely *a priori* that the molecule forms pores. Second, since glycophorin contains no sulfhydryl groups (Marchesi, 1979), it cannot be the target of PCMBS action in closing the water pores, and therefore presumably cannot be the pore former.[‡] This leaves band 3 and band 4.5. As we shall see, there are problems with both of them.

Band 3 is at first glance (and even second glance) a very attractive candidate for the water pore. It is the major protein component of the

---

[*] I refer here to glycophorin A, which comprises about 75% of the PAS-staining protein (Marchesi, 1979).

[†] It should be appreciated that since membrane proteins are identified and picked up on gels, stained with agents such as Coomassie blue or PAS, the possibility cannot be precluded that there exist undiscovered proteins in high copy number that have thus far escaped detection. My biochemist friends (who are few in number) tell me that this is unlikely.

[‡] It is conceivable that the water pore is formed by a multimer of glycophorin held together by glycophorin attachment to a network of submembranous proteins, which in turn have sulfhydryl groups. The reaction of these sulfhydryl groups with PCMBS could lead to disruption of the network and subsequent disaggregation of the multimer forming the pore. I am indebted to Dr. Vincent Marchesi for pointing out this possibility to me.

red cell membrane and contains the anion transporting activity (see Cabantchik et al., 1978). Although there are other proteins besides the anion transporter in band 3, they are minor components in small copy number (Cabantchik et al., 1978; Knauf et al., 1974). Of particular relevance to its role in water transport is the demonstration by Sha'afi and Feinstein (1977) that PCMB can be shown to bind specifically to band 3 under the conditions in which it inhibits water transport. To demonstrate this, the nonspecific binding of PCMB to other proteins in the membrane was first blocked by pre-incubating the cells with N-ethylmaleimide (NEM), iodoacetamide (IAM), or mersalyl, reagents that react with many of the sulfhydryl groups in the membrane but do not inhibit water transport or interfere with the inhibitory action of PCMB (Sha'afi and Feinstein, 1977). [Similar results were obtained by Brown et al. (1975) with 5,5'-dithiobis-2-nitrobenzoic acid (DTNB), but the claim that it is an effective inhibitor of water transport (Naccache and Sha'afi, 1974) is disputed; Macey (1984) and Brahm (1982) report no inhibition, and Levitt and Mlekoday (1983) and Benga et al. (1983) find only a small inhibition.] Additional evidence implicating band 3 with the water pores comes from freeze etch experiments. Pinto da Silva (1973) found that sublimation of water molecules at $-100°C$ occurs primarily through regions of erythrocyte ghost membranes that contain particle aggregates, and he suggested that this preferential pathway for the passage of water molecules through the plasma membrane might be the water pores. Subsequent experiments by Pinto da Silva and Nicolson (1974) demonstrated that these particle aggregates contain the band 3 protein.

There are nonetheless major objections to the proposition that water and anions share a common permeation pathway through the band 3 protein. (1) It is difficult to envision how DIDS (4,4'-diisothiocyano-2,2'-stilbene disulfonate) can completely inhibit anion transport without affecting water permeability (Benga et al., 1983; Solomon et al., 1983; Macey, 1984). (2) Even more difficult to reconcile with a common pathway is the ability of PCMBS to completely block pore-mediated water transport without affecting anion permeability (Knauf and Rothstein, 1971; Knauf, 1979; Knauf et al., 1983). [Solomon et al. (1983) attempt to rationalize these data, but their arguments are not very convincing (see Macey's (1984) rebuttal of their points).] (3) A further problem in trying to argue that water and anions permeate through the same pathway is the existence of red cells, such as those of the chicken, that have a facilitated anion transport system comparable to that of the human red cell [i.e., the same anion turnover number and roughly the same copy number per cell (Jay, 1983)], and yet do not have water pores, as judged by the low value of P_f, $P_f = P_{d_w}$, the high Q_{10} for water permeability, and the failure of PCMBS to inhibit water permeability (Farmer and Macey, 1970; Blum and Forster, 1970; Brahm and Wieth, 1977).

These objections to the proposal that the water pores are associated with the anion exchange protein (band 3) are circumvented if it is assumed that the water and anion transport systems are independent. For instance, there is considerable evidence that the band 3 protein may exist as a dimer in the membrane (see Knauf, 1979), but it is not yet clear whether the monomer or the dimer is the functional anion transporting unit. Whichever one it is, the other could be the water pore. For example, if the monomer functioned as the anion transporter, the water pore could be formed between the two monomers. Alternatively, the pore could be formed by the association in the membrane of the band 3 protein with another protein (e.g., glycophorin), or as a leakage path between its intramembranous part and the adjacent surrounding lipid. Suggestions such as these have been made by several investigators (e.g., Brown et al., 1975; Brahm, 1982). These possibilities account for the water pores' association with band 3, and yet are consistent with the independent inhibition of water and anion transport by agents such as DIDS and PCMBS. They also offer an explanation for the absence of water pores in red cells (such as those of the chicken) that still have an active band 3-facilitated anion transport system. Namely, the association of the band 3 protein with itself or other elements in the membrane may not be identical in all cells, because of subtle variations in its structure (or in that of the proteins with which it interacts) unrelated to its anion transporting activity. As I pointed out in the previous chapter, permeation of water through channels in most cells, and red cells are included among these, must be incidental to other functions of the protein forming the channel. The possibilities mentioned here are in keeping with that interpretation.

What is the prospect that band 4.5, the glucose transporting system (Wheeler and Hinkle, 1985), is the water pore? The most serious objection to this possibility is that in adult erythrocytes, the glucose transport system exists at high levels of activity (and therefore presumably in high copy number) only in those of primates (Laris, 1958), whereas many other erythrocytes have water pores. For example, the beef erythrocyte has a water permeability comparable to that of the human (Farmer and Macey, 1970), but virtually no facilitated glucose transport system (Laris, 1958).

In sum, the most likely candidate for the water pore is the band 3 protein. The pore it creates for water, possibly in association with other proteins or with lipids, is an independent entity, providing a route for water flow independent of facilitated anion exchange.

F. WHAT ABOUT OTHER PORES?

One further issue needs to be addressed in the context of counting pores: the urea transport system. Urea flux across the red cell membrane is frequently described as occurring by facilitated transport (e.g., Mayrand and Levitt, 1983; Brahm, 1983b). This characterization, however, is based on findings of saturation kinetics and competitive inhibition, phenomena equally well explained by pores having saturable, molecular-specific binding sites. [The gramicidin A pore, for example, manifests these characteristics with respect to univalent cations (see Hladky and Haydon, 1984)]. Taking this latter point of view, let us make an estimate of the number of these pores required to account for the urea permeability of the red cell membrane.

Given that $P_{d_{\text{urea}}}$ in the limit of zero urea concentration is 1.16×10^{-3} cm/s and that $K_m = 218 \, \text{m}M$ (Mayrand and Levitt, 1983), the unidirectional urea flux per cell is given by:

$$\Phi_{\text{urea}} = P_{d_{\text{urea}}} A c_{\text{urea}} = \frac{P_o A c_{\text{urea}}}{1 + \dfrac{c_{\text{urea}}}{K_m}} \qquad (10\text{-}7)$$

where,

$$P_o = P_{d_{\text{urea}}} \left(1 + \frac{c_{\text{urea}}}{K_m} \right) = 1.16 \times 10^{-3} \text{ cm/s} \qquad (10\text{-}8)$$

$$A = 137 \ \mu m^2 \qquad (10\text{-}9)$$

The maximum unidirectional urea flux *per cell,* $\Phi_{\text{urea}}^{\text{max}}$, is obtained in the limit of $c_{\text{urea}} \to \infty$, and becomes, from eqs.(10-7) and (10-8):

$$\Phi_{\text{urea}}^{\text{max}} \approx 3.5 \times 10^{-13} \text{ moles/s/cell} \approx$$
$$2 \times 10^{11} \text{ molecules/s/cell} \qquad (10\text{-}10)*$$

The magnitude of this number by itself strongly suggests that urea transport occurs through pores and not by facilitated diffusion. The largest turnover number, by far, for a facilitated transport system is that of the anion exchange system of the red cell at 38°C: $\approx 5 \times 10^4$ ions/s/site (Brahm, 1977); most other facilitated transporters have turnover numbers of 10^2 molecules/s/site or less. Even if this maximum value applies to the urea transport system, there have to be [from eq.(10-10)] approximately 10^7 transport sites per cell—more than the total number of identifiable proteins in the red cell membrane.

What is a reasonable number for the maximum unidirectional flux *per pore* ($M_{\text{urea}}^{\text{max}}$) for the urea-transporting system? Unlike the case of water, where results from the gramicidin A pore can be directly applied to the water pores, there is no model pore with selectivity and saturation kinetics comparable to those of the urea pores. Our esti-

*The value obtained by Brahm (1983b) is approximately one third of this.

mate, therefore, must rest on more tenuous grounds. Nevertheless, a reasonable inference can again be made from the gramicidin A pore, this time based on its Na^+ permeability. The K_m of 310 mM for Na^+ in the gramicidin A pore (Finkelstein and Andersen, 1981) is comparable to the K_m of 218 mM for urea in the urea pore (Mayrand and Levitt, 1983), and therefore the maximum unidirectional flux (at zero voltage) of the former is of some relevance to that of the latter. For sodium in the gramicidin A pore, it is 2.3×10^6 ions/s/pore (Finkelstein and Andersen, 1981). This value for urea in the urea pore means [from eq.(10-10)] that there are approximately 10^5 urea pores per cell.

M_{urea}^{max} could obviously be smaller than 2.3×10^6 molecules/s/pore, in which case the number of urea pores per cell would be even greater than 10^5. This creates two problems. First, playing the same numbers game with these pores as played with the water pores, we are forced to conclude that in addition to containing the facilitated anion transport system and the water pores, the band 3 protein also contains the urea pores. Second, since it is hardly conceivable that water does not also go through the urea pores, then if they are as numerous as the water pores, a significant fraction of the water permeability pathway must be shared with urea—a conclusion we previously rejected.

M_{urea}^{max} could also be greater than 2.3×10^6 molecules/s/pore,* thereby reducing the number of required pores and making the protein responsible for them a minor component of the plasma membrane; the water flux through the urea pores would then be small compared to that through the water pores and would be masked by the latter. We

---

*Given the value of K_m, the only way that M_{urea}^{max} can be significantly greater is for both the entry and exit rate constants to be proportionally larger than in the gramicidin A pore [see eq.(24) of Finkelstein and Andersen (1981)]. There is no difficulty in envisioning a larger exit rate constant. The entry rate constant for the gramicidin A pore, on the other hand, is diffusion limited (Andersen, 1983a,b), with a capture radius of approximately 0.2 Å (Andersen, 1983a). If the entrance to the urea pore is 4 Å in radius, instead of the 2 Å for the gramicidin A pore, the capture radius for urea would become approximately 2 Å (instead of 0.2 Å), and M_{urea}^{max} could then be 10-fold greater than the estimate based on $M_{Na^+}^{max}$ for the gramicidin A pore.

argued previously that there are no special proteins in the red cell membrane devoted to water pores, since the cell does not "need" a high water permeability, and that therefore the water pores arise incidentally to other functions of the protein that forms them. The same consideration, however, is not applicable to urea pores. Macey (1984) has pointed out a physiological role for a rapid urea transport system in the red cell, and therefore we should not be surprised to find specific proteins in the cell membrane devoted to this function.

A similar calculation to the one made for urea can be made for ethylene glycol, also using the data of Mayrand and Levitt (1983). An estimate of approximately 2.5×10^4 pores/cell is obtained. These pores are probably also responsible for glycerol transport [Mayrand and Levitt (1983)] and, like the urea pores, could be formed by a minor membrane component.

G. CONCLUSION

The body of evidence reviewed and summarized in this chapter indicates that the major permeation routes across the red cell membrane for water, urea, ethylene glycol, and anions are independent. An estimate of the number of water pores per cell required to account for the water permeability strongly suggests (along with other information) that the water pores are associated with the anion transporting system (band 3). One can argue (as we have) that there are many fewer urea and ethylene glycol pores, but it is conceivable that they too are as numerous. If that is so, the simplest way to accommodate all these transport systems in the plasma membrane is for them to share the same pathway through the band 3 protein, which is the point of view taken by Solomon et al. (1983). I tend to agree with Macey (1984), however, that this is unlikely, as too many *ad hoc* arguments are required to explain away all the data, reviewed in this chapter, that are inconsistent with such a model. The role of band 3 in water, urea, and ethylene glycol transport may eventually be resolved from transport measurements on protein reconstituted into lipid bilayer membranes;[†] the recent cloning and sequencing of the band 3 protein (Kopito and Lodish, 1985) should also be of considerable aid in clarifying this issue.

[†] The conductance effects observed by Benz et al. (1984) in planar lipid bilayer membranes into which band 3 protein was incorporated are uninterpretable in terms of the known permeability characteristics of the red cell membrane, and their attempts to relate them to water permeability seem far-fetched. The claim by Carruthers and Melchior (1983) that the water permeability of lipid vesicles is substantially increased when erythrocyte integral membrane proteins are incorporated into them is difficult to credit, given that the water permeabilities of their unmodified egg PC vesicles are more than an order of magnitude smaller than those obtained on egg PC bilayers (be they vesicles or planar membranes) by all other investigators, none of whom the authors cite.

11

Epithelia:
Antidiuretic Hormone (ADH)-Induced
Water Permeability

Few aspects of transport physiology are as relevant to the overall physiological state of the organism, and as fascinating, as the functioning of epithelia. The fluxes of water, nonelectrolytes, and salts across these tissues are governed by an emporium of ion- and solute-specific transport systems, and to hear renal physiologists talk, their integration into tissue or organ function rivals in complexity that of the central nervous system. Epithelia, therefore, have justly deserved the attention of the general physiologist. For the biophysicist or membrane physiologist, however, interested in the characteristics and mechanisms of action of specific transport systems, epithelia are a nightmare. Instead of examining transport across a single plasma membrane, as is done with the red cell or squid giant axon, the investigator is confronted by transport across at least two plasma membranes, in series with each other and with additional barriers such as cell cytoplasm and basement membranes. Moreover, heterogeneity of cell type within the epithelial sheet and the presence of paracellular transport routes further confound the interpretation of experimental data. In no area of epithelial transport are these problems more exacerbated than in that of water transport. My own minimal excursions into this field are characterized by a love-hate relationship, with emphasis on the latter; in fact, the writing of this chapter represents a partial breaking of my recent vow of total abstinence from epithelia.

The main focus of this chapter is on the nature of the water permeability pathway induced by antidiuretic hormone (ADH) in epithelia responsive to its action. As problems in epithelia go, this is a relatively simple one. Yet, as we shall see, there are formidable difficulties in trying to obtain and interpret water permeability data in order to establish one simple fact: the value of P_f/P_{d_w}. Before we take up this topic, a brief general overview of the water permeability characteristics of epithelia is given, so as to set in context the action of ADH on toad urinary bladder and related epithelia responsive to this hormone.

A. GENERAL WATER PERMEABILITY CHARACTERISTICS OF EPITHELIA

Whether the primary route of transepithelial water flow is transcellular or paracellular has been a major issue among epitheliologists for some time. Figure 11-1 is a schematic diagram of an epithelium illustrating the possible routes of water flow across the epithelial cell layer. (The subepithelial structures that must ultimately be crossed are omitted from the diagram.) Under physiological conditions, water generally flows from the mucosal (also frequently called the luminal or apical) side of the tissue to the serosal (or basal or basolateral) side. For the transcellular pathway, water crosses the mucosal membrane, enters the cell, and then exits across either the basal or lateral membranes. (In the latter instance, water enters the lateral intercellular space, making the last part of the transcellular route paracellular; this is not, however, what is at issue with respect to the question at hand, although these spaces have played an important and controversial role in theories of fluid transport.) For the paracellular route, water crosses the mucosal surface through the (misleadingly named) tight junctions into the lateral intercellular space and then out into the serosal solution, without crossing any cell membranes.

The question of the route of water transport has been of particular interest with respect to so-called "leaky" epithelia, such as gall bladder or the proximal tubule of the kidney. These epithelia derive their name from their large passive ion permeability, as manifested by very low transepithelial electrical resistances (Frömter and Diamond, 1972). Coincident with the large ion (and nonelectrolyte) permeability there is often [but not always; e.g., thick ascending limb of the renal tubule (Hebert et al., 1981)] a very high water permeability, making the epithelium also "leaky" to water. Since it is established that passive ion (and probably nonelectrolyte) flux across leaky epithelia occurs primarily paracellularly through the tight junctions (e.g., Frömter, 1972; Frömter and Diamond, 1972; Wright and Pietras, 1974; Berry and Boulpaep, 1975), it is often suggested that this is also the principal route for water transport. The controversy surrounding this question has been recently reviewed by Berry (1983) and Spring (1983), and the weight of evidence now appears to be on

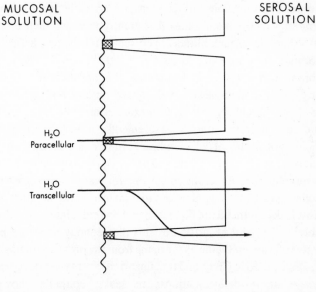

Figure 11-1. Schematic diagram of an epithelium illustrating the transcellular and paracellular pathways for water. The sub-epithelial structures of the tissue, in series with this epithelial layer, are not shown.

the side of the transcellular, rather than the paracellular, pathway as the primary route for water transport.

The major arguments for this point of view are twofold. The first is a theoretical argument based on the fraction of total mucosal area occupied by the tight junctions and an estimate of what is a reasonable value for P_f per junction. The upshot of the calculations is that only a trivial fraction of the water flow in leaky epithelia could occur through the junctions, and that consequently virtually all of the flow is transcellular (see Spring, 1983). The second argument arises from direct measurements of the water permeabilities of the luminal and basolateral membranes. In *Necturus* gallbladder, where they were first measured, the P_f's of the luminal and basolateral membranes are 5.5×10^{-2} cm/s and 0.12 cm/s, respectively (Persson and Spring, 1982). These values are unusually large, because they are expressed in terms of the ostensible surface areas, which neglects the microvilli or infoldings of the membranes that greatly increase their effective areas. When the permeabilities are corrected for the true areas of the membranes, they turn out to be 6.8×10^{-3} cm/s for the luminal membrane and 4.6×10^{-3} cm/s for the basolateral membrane (Spring, 1983). Comparable values are obtained in other leaky tissues (e.g., Fischbarg, 1982; Gonzalez et al., 1982).

The results from direct measurements of water permeability coefficients of epithelial cell membranes have come as somewhat of a revelation to workers in the field and have been decisive in establishing the transcellular route as the major one for water flow across leaky epithelia. Spring (1983) points out that measured transepithelial P_f's for most leaky epithelia are far less than those calculated from the cell membrane P_f's, and he attributes this discrepancy to errors in the estimates of transepithelial P_f's, arising from the presence of unstirred layers. [We noted in Chapter 3 that unstirred layers were not generally an issue in P_f determinations; in leaky epithelia, however, water permeabilities and flow rates are so large that they can become a serious problem (e.g., Diamond, 1979).] This is illustrative of my comments at the beginning of the chapter on the difficulties in obtaining and interpreting transepithelial permeability data. The development of methodologies for direct measurements of membrane permeabilities is an important advance that should have a growing impact on epithelial physiology.

The magnitudes of membrane P_f's (corrected for the true membrane areas) are not out of line with values obtained on unmodified lipid bilayers and might therefore simply be a reflection of the intrinsic water permeabilities of epithelial membrane bilayers. In other words, there is no necessity to invoke, *a priori,* water pores to account for the water permeabilities of these membranes. On the other hand, the recent finding by Wittembury et al. (1984) that PCMBS causes both a 4-fold reduction in P_f of the basolateral plasma membrane of the rabbit's proximal straight tubules and an increase in its Q_{10} indicates the presence of a significant porous route for water transport across that membrane. In general, however, the mechanism of water permeation across the plasma membranes of leaky epithelia remains an open question requiring further investigation.

B. TOAD URINARY BLADDER

Unlike the epithelia discussed in the preceeding section, the urinary bladder of the toad is a "tight" epithelium with respect to its ion and nonelectrolyte permeability, and it also has a low water permeability in the basal state. In response to ADH, however, an enormous increase of 50-fold or so occurs in its water permeability; the nature of this response is the subject of this section. The hydroosmotic effect of ADH* is not confined to the toad urinary bladder; other notable responsive tissues are frog urinary bladder, mammalian cortical collecting kidney tubules, and frog and toad skin.

---

*Besides its effect on water permeability, ADH can also affect urea and Na^+ permeability. The relation, or lack thereof, of these latter actions to its effect on water permeability will be mentioned in the course of the discussion.

CHAPTER 11

The last two were the objects of study in the classic paper by Koefoed-Johnsen and Ussing (1953), which was among the first to interpret values of P_f/P_{d_w} greater than 1 in terms of pores and pore radii. The hydroosmotic response of epithelia to ADH stimulation ranks with water transport across the red cell membrane as the most venerable and extensively investigated topic in the field of water permeation through cell membranes. Almost all of the issues relevant to the water permeability of bilayers and pores arise in this system, so it is particularly appropriate to close this monograph with a chapter that reviews the state of this subject. The toad urinary bladder is chosen as the focus of this discussion, both because it is the ADH-responsive tissue that has received the most attention, and because it is a preparation with which I have had some direct experience.

The Basal State

In its unstimulated state, the water permeability of the bladder lies at the low end of the spectrum for plasma membranes. Transepithelial values for P_f and P_{d_w} are approximately 2.5×10^{-4} cm/s and 0.8×10^{-4} cm/s, respectively (Levine et al., 1984b). The basal state has not been extensively investigated, and accurate values of P_f are especially difficult to obtain, because of the small osmotically induced flow rates. I do not feel, therefore, that particular stress can be placed on the basal value of P_f/P_{d_w}. It is also not clear for this tight epithelium whether transepithelial water flow is primarily transcellular or paracellular, as the arguments for transcellular flow in leaky epithelia are not applicable to an epithelium with such a low water permeability [see Spring's (1983) discussion of this point]. What is clear is that the water permeability of the luminal (mucosal) membrane is much smaller than that of the basolateral (serosal) membrane and that therefore the low permeability of the tissue is attributable to the permeability properties of the former. The clearest illustration of this is the failure of the epithelial cells to swell when the mucosal medium is diluted, in contrast to their rapid swelling in response to dilution of the serosal medium (Di Bona et al., 1969).

The ADH-Stimulated Bladder

Mechanism of Action. The hormonal effect of ADH is mediated through a second messenger, 3',5'-cyclic AMP (see Handler and Orloff, 1973). Although ADH must be added to the serosal side of the bladder (where it interacts with receptors on the basolateral membrane to stimulate an adenylate cyclase), its effect on water permeability (as well as on urea and Na^+ permeability) occurs at the mucosal membrane of the granular cells, which now swell when the mucosal medium is diluted (Di Bona et al., 1969). The increased transepithelial water permeability is thus a consequence of the increased water permeability of the mucosal plasma membrane. How is this change in permeability brought about? In particular, what is the mechanism of water permeation across the ADH-stimulated mucosal membrane? Most of the attempts to answer this latter question have centered on the value of P_f/P_{d_w}.* The theoretical and experimental efforts devoted to the determination and interpretation of this quantity have an interesting history that is useful to recall, as it illustrates so many of the issues discussed in preceeding chapters.

In the first studies dealing with this problem, Koefoed-Johnsen and Ussing (1953) found that ADH stimulation of frog or toad skin resulted in large increases in P_f but relatively modest increases in P_{d_w}; subsequent experiments with toad urinary bladder confirmed this finding in that tissue as well (Hays and Leaf, 1962). Since, as discussed in Chapter 3, the value of P_f/P_{d_w} in a porous membrane increases with pore radius, it was natural to conclude that water traverses the unstimulated bladder (or skin) through pores, and that the action of ADH is either to increase their radius or to open new ones of larger radius (Koefoed-Johnsen and Ussing, 1953; Hays and Leaf, 1962); indeed, pore radii were calculated by Andersen and Ussing (1957) from the magnitude of P_f/P_{d_w}.

---

*Measurements of solvent drag effects or of the Q_{10} for P_{d_w} have foundered on unstirred layer problems [see Hays's (1972) critique]. Interpretations of the Q_{10} for P_f are complicated by possible multiple effects of temperature on water permeability, as discussed in Chapter 9, but ingenious approaches have been employed, with some success, to deal with these problems (e.g., Eggena, 1972; Kachadorian et al., 1979).

When Dainty (1963) drew attention to the presence of unstirred layers around membranes and to their potential for distorting P_f/P_{d_w} values, because of their tendency to cause underestimations of P_{d_w} (see Chapter 3), doubts arose over previous determinations of P_f/P_{d_w} in ADH-stimulated tissues. Dainty and House (1966) found that stirring *did* affect P_{d_w} measurements in frog skin and therefore affected the value of P_f/P_{d_w}. However, at maximal stirring rates, they found that ADH treatment, although sizably increasing P_f, caused no significant increase in P_{d_w}, thereby seeming to confirm Koefoed-Johnsen and Ussing's (1953) original observation. They recognized, however, that unstirred layers, inaccessible to outside mechanical stirring, might be present within the layers of the skin itself, and were therefore hesitant to state an exact value for P_f/P_{d_w} or even to conclude definitively that P_f/P_{d_w} was greater than 1 in ADH-stimulated frog skin. Hays and Franki (1970) found a dramatic decline with stirring rate in the value of P_f/P_{d_w} for ADH-stimulated toad urinary bladders. They obtained a minimum value of approximately 40, but also recognized, and demonstrated, the existence of diffusion barriers within the tissue, and therefore could reach no conclusion concerning the true value of P_f/P_{d_w}, which they felt might in fact be 1.

These results, plus those of water permeability measurements on lipid bilayer membranes, shifted opinion to the view that the water permeability of the mucosal plasma membrane might be understood simply in terms of its bilayer structure. When it was realized that the magnitude of P_f (and P_{d_w}) of lipid bilayer membranes is a function of their lipid composition and can take on the wide range of values found in cell membranes, it was proposed that water traverses the mucosal membrane through its bilayer proper and that the true value of P_f/P_{d_w} is 1. The ADH-stimulated rise in water permeability was assumed to result from an increase in bilayer fluidity, brought about by a hormone-induced modification of bilayer composition or structure (Schafer et al., 1974; Pietras and Wright, 1975).

This explanation for ADH's effect on water permeability is at first sight an attractive one, ostensibly consistent with water permeability data from lipid bilayers. Finkelstein (1976b) pointed out, however, a major difference between the effects of lipid composition and fluidity on bilayer permeability and the effects of ADH on bladder permeability. Namely, modifications of bilayer composition and structure that increase water permeability also produce comparable increases in nonelectrolyte permeabilities (Finkelstein, 1976a; see also Table 6-1), whereas the 50-fold increase in the water permeability of the toad urinary bladder that is induced by ADH is accompanied by negligible increases in permeability to most nonelectrolytes (Pietras and Wright, 1975). Table 11-1 compares the values of $P_f/P_{d_{solute}}$ (where the solute is n-butyramide, isobutyramide, 1,4-butanediol, and 1,6-hexanediol) in three different lipid bilayers to their values in ADH-stimulated bladders. As we noted in Chapter 6, the value of $P_f/P_{d_{solute}}$ for a given solute is reasonably independent of bilayer composition; in contrast, its value for the stimulated bladder is much greater. For example, $P_f/P_{d_{n\text{-butyramide}}}$ is over a thousand-fold greater in the ADH-stimulated bladder than in lipid bilayers. Unless the bilayer of the mucosal membrane of the toad bladder possesses unprecedented properties, the obvious conclusion from these data is that the major route for water permeation through that membrane is *not* through its bilayer proper, but rather is through pores.

The Value of P_f/P_{d_w}. It has been recognized for some time that if water crosses the ADH-stimulated bladder through pores, these pores must be quite narrow, since the permeability coefficients of most small solutes are not substantially increased by ADH (Leaf and Hays, 1962). It was originally believed that urea and small amides (such as acetamide) could pass through these pores, since their permeability coefficients also greatly increase following ADH treatment, but this conclusion no longer appears to be correct. As with the red cell membrane, phloretin inhibits urea and amide permeability without affecting water permeability (Levine et al., 1973). Moreover, certain general anesthetics inhibit water permeability without affecting urea and amide permeability (Levine et al., 1976). The ability to decouple these permeabilities implies that water and urea pathways

TABLE 11-1. *Ratios of P_f to $P_{d_{solute}}$ for Lipid Bilayers and ADH-Stimulated Toad Urinary Bladder*

Membrane	$P_f/P_{d_{n\text{-butyramide}}}$	$P_f/P_{d_{isobutyramide}}$	$P_f/P_{d_{1,4\text{ butanediol}}}$	$P_f/P_{d_{1,6\text{ hexanediol}}}$
Sph/chol, 14.5°	4.1	—	—	—
Sph/chol, 25°	2.8	6.9	27	1.8
PC/chol, 25°	1.9	2.9	29	2.6
ADH-stimulated bladder	6,000	14,000	19,000	4,600

The values for the sphingomyelin/cholesterol membranes (Sph/chol) and phosphatidylcholine/cholesterol membranes (PC/chol) are calculated from the data in Table 6-1. The values for the ADH-stimulated bladder are calculated from the data of Pietras and Wright (1975), and a value of 5×10^{-2} cm/s for P_f (Levine et al., 1984a; see the second footnote on page 198).

are separate in the toad urinary bladder.* In addition, ADH increases water permeability alone in cortical collecting tubules, without affecting urea permeability (Grantham and Berg, 1966). It therefore appears that the ADH-induced pores, like the gramicidin A pores and the water pores of the red cell membrane, are very narrow in at least some region, and that water transport through that region must be single-file. A correct measurement of P_f/P_{d_w} should then reveal the number of water molecules in single-file array within the pore [see eq.(4-17)] and thereby help characterize the pore structure.

Levine et al. (1984a,b) succeeded in making this measurement on toad urinary bladders. Their approach was to first determine the water permeability of the barriers, both within and outside the tissue, in series with the mucosal membrane, so that measured values of P_{d_w} could be corrected for what are in effect unstirred layers. In fact, once the magnitude of this series barrier correction was known, experiments could be designed, using submaximal doses of ADH or 8-bromo cyclic AMP, to minimize the relative size of this correction to measured values.

In their first paper, Levine et al. (1984a) determined the value of P_{d_w} for the series barriers by increasing the water permeability of the mucosal membrane to such an extent (with maximum doses of either ADH or amphotericin B) that it no longer offered a significant resistance to water diffusion; hence, the measured value of P_{d_w} was essentially that of the series barriers. (It turned out that with the effective stirring they employed, virtually all of the series barrier to the diffusion of water resided within the tissue.) In their second paper, Levine et al. (1984b) used the series barrier data in conjunction with values of P_{d_w} measured in bladders stimulated with low doses of ADH or 8-bromo cyclic AMP to calculate P_{d_w} for the mucosal membrane. Combined with the simultaneous determination of P_f, this then gave the true value of P_f/P_{d_w} for the hormonally stimulated mucosal membrane. As a check on their methodology, they also determined

---

*The ADH-activated Na^+ permeability is probably attributable to still another pathway (Ferguson and Twite, 1974; Carvounis et al., 1979).

P_f/P_{d_w} for mucosal membranes stimulated (submaximally) with amphotericin B.

The results of their study are shown in Figure 11-2. The major finding was that P_f/P_{d_w} for the ADH- or 8-bromo cyclic AMP-stimulated mucosal membrane is large. In particular,

$$\frac{P_f}{P_{d_w}} \approx 17 \qquad \text{(mucosal membrane of stimulated bladder)} \qquad (11\text{-}1)$$

[Their value of 3.7 for P_f/P_{d_w} for the amphotericin B-stimulated mucosal membrane was in reasonable agreement with the value of 3.0 for lipid bilayer membranes, thereby lending credence to their result for the ADH-induced (or 8-bromo cyclic AMP-induced) water

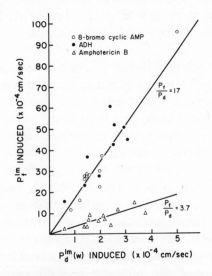

Figure 11-2. Graphs of P_f vs. P_{d_w} for the mucosal membrane of toad urinary bladders stimulated with 8-bromo cyclic AMP (○), ADH (●), or amphotericin B (△). (The basal values have already been subtracted from the experimentally measured values, so that stimulus-induced values are being plotted.) [From Levine et al. (1984b).]

permeation pathway (eq.(11-1).] Parisi and Bourguet (1983), using a different approach for dealing with unstirred layers, also obtained a large value for P_f/P_{d_w} in the ADH-stimulated frog urinary bladder. They reported a value of 9, but felt that a value of twice this was also consistent with their data.*

The Meaning of $P_f/P_{d_w} \approx 17$. If P_f/P_{d_w} is assigned a value of 17 in eq.(3-6), a radius of approximately 15Å is obtained. This is clearly inconsistent with the impermeability of ADH-induced pores to small nonelectrolytes, including urea. Given the implied single-file nature of water movement through these pores mentioned earlier, eq.(3-6) was obviously, *a priori,* inappropriate. Equation (4-17) offers a more logical interpretation of the value of P_f/P_{d_w}—namely, the pore contains approximately 17 water molecules in single-file array. This number of water molecules can reasonably be accommodated in a pore of 2 Å radius and 50 Å length, the latter dimension being compatible with its spanning the lipid bilayer of the plasma membrane. Such a model is consistent both with the value of P_f/P_{d_w} and with the pore's impermeability to small nonelectrolytes. Another constraint, however, should be considered in assessing pore models and the value of P_f/P_{d_w}.

Electron microscopic studies of ADH-stimulated bladders in conjunction with permeability measurements provide compelling evidence that ADH-induced water permeability results from fusion with the mucosal membrane of cytoplasmic tubular vesicles, approximately 2 μm in length and 0.1 μm in diameter (Kachadorian et al., 1981), containing intramembranous particles arranged as "aggregates" (Kachadorian et al., 1977a; Muller et al., 1980; Wade, 1980). The implication is that the water-permeable pores are present in the vesicles' membranes and that through the fusion process they

---

*It is unclear what the true value of $P_f P_{d_w}$ is for ADH-stimulated cortical collecting tubules of mammalian kidneys. The P_{d_w} measurements (Schafer et al., 1974; Al-Zahid et al., 1977; Hebert and Andreoli, 1980) are so dominated by series barrier contributions that it is impossible to establish the value of P_{d_w} for the mucosal membrane and hence the true value of P_f/P_{d_w}.

are functionally inserted into the mucosal membrane. Whether most of the pores remain in the walls of the tubular vesicles or migrate out onto the surface of the mucosal membrane turns out to be an important issue which we shall consider shortly. Some of the aggregates certainly migrate out onto the surface, and since the correlation between the ADH-induced water permeability and the number of aggregates appearing on the luminal membrane is excellent (Kachadorian et al., 1975, 1977a; Levine and Kachadorian, 1981), it is generally assumed that the increased water permeability is through the area delineated by these surface aggregates.* The correlation would still exist, however, if the water permeability is through the entire tubular membrane, which is lined by aggregates and is accessible to the mucosal bath through the area of fusion (Muller et al., 1980). (It is also possible, but unlikely, that the aggregates have nothing directly to do with water permeability but are merely identifiable markers of pore-containing vesicles. Thus, when these vesicles fuse with the membrane, both the water pores and the aggregates are inserted into the mucosal membrane, but only the latter are identifiable in electron micrographs of freeze-fracture replicas.)

Let us assume for the moment that the area occupied by the aggregates on the mucosal surface of the bladder contains all the water pores. A fully stimulated bladder has a P_f for the mucosal membrane of $\sim 5 \times 10^{-2}$ cm/s (Levine et al., 1984a)[†] and contains $\sim 10^8$ aggregates/cm^2 (Kachadorian et al., 1977b). The p_f of each aggregate is therefore,

$$(p_f)_{\text{aggregate}} = \frac{5 \times 10^{-2}}{10^8} = 5 \times 10^{-10} \text{ cm}^3/\text{s} \qquad (11\text{-}2)$$

*Using this assumption, Kachadorian et al. (1979) normalized water flow for the number of aggregates and thereby obtained a small Q_{10} for P_f, from which they concluded that water movement through the mucosal membrane of ADH-stimulated bladders is through pores.

[†]The value of $\sim 2 \times 10^{-2}$ cm/s found by Levine et al. (1984a) for P_f of the entire tissue has to be increased by more than a factor of 2, because of series barriers to water flow (Levine and Kachadorian, 1981).

Taking the p_f of the gramicidin A pore ($\sim 3 \times 10^{-14}$ cm³/s) as an upper limit for that of the bladder's single-file water pores (which are probably twice the length of the gramicidin A pore) and combining it with the value in eq.(11-2) for $(p_f)_{aggregate}$, we have:

$$\frac{5 \times 10^{-10}}{3 \times 10^{-14}} \approx 2 \times 10^4 \text{ pores/aggregate} \qquad (11\text{-}3)$$

This number of single-file pores must be placed in an area of approximately 10^{-2} μm^2, the area occupied by an aggregate [~ 0.1 μm diameter (Kachadorian et al., 1977b)]. The area available for each pore is therefore $\sim 10^6/(2 \times 10^4) = 50$ Å², or in other words, the center to center distance between the pores is approximately 7 Å, only slightly larger than their 4 Å internal diameter. This leaves only about 2 Å for the walls of the pores, which is incompatible with any realistic molecular structure. [Wade (1980) makes a similar calculation for his "sieve" model of the pore.] It therefore appears that not enough area is provided by the aggregates to accommodate the number of single-file pores required to account for the ADH-induced P_f.

To circumvent this problem, Levine et al. (1984b) proposed a "showerhead" model for the pore (Fig. 11-3), consisting of a long stem with a large radius (e.g., length 50 Å, radius 20 Å) ending in a

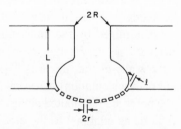

Figure 11-3. Shower-head model for ADH-induced water pore. A large number of short (length *l*), small radius (*r*) single-file regions are in parallel with each other and in series with a long (length *L*), large radius (*R*) tube. [From Levine et al. (1984b).]

wide circular cap (e.g., radius 200 Å) containing many ($\sim$ 500) short holes of 2Å radius. Such a structure would hold back small solutes (because of the short, 2 Å-radius holes in the cap) but have a large value for P_f/P_{d_w} (because of the long, large-radius stem in series with the cap of small-radius holes). [See Levine et al. (1984b) for the details of the calculation.] Six of these structures, which can be accommodated in the area occupied by an aggregate, have the water permeability attributed to it [eq.(11-2)] (Levine et al., 1984b). This model is formally similar to the dual barrier hypothesis proposed years ago for the toad skin (Andersen and Ussing, 1957) and toad urinary bladder (Leaf and Hays, 1962; Lichtenstein and Leaf, 1966) to resolve the seeming paradox of their simultaneously displaying both a large value for P_f/P_{d_w} and an impermeability to small solutes. The idea was that the tissue contains two barriers in series—one of fine pores that sieve small solutes, and the other an ADH-sensitive region that develops wide pores in response to the hormone.

There is an alternative explanation for the large value of P_f/P_{d_w}, which does not embody any specific features for the pores other than the constraint that they contain a narrow region ($\sim$ 2 Å radius) to prevent the permeation of small nonelectrolytes. At the heart of this explanation is the hypothesis that water permeability is attributable not only to pores that have migrated out onto the mucosal surface, but also to pores that line the tubular membrane throughout its length and are accessible to the mucosal bath through the area of fusion. It turns out that because of the long ($\sim$ 2 μm) diffusion path for water down the tubule, P_f/P_{d_w} is determined by the dimensions of the tubular vesicle and can therefore be very large, even if P_f/P_{d_w} for the individual pores in the walls of the tubule is 1 [see Levine et al. (1984b) for the details of the calculation]. An attractive feature of this model is that since the pores are distributed along the entire length of the tubule, rather than being confined to the area occupied by the aggregates on the mucosal surface, there is about a 10-fold larger area available for them. Consequently, there is no difficulty in packing enough pores into this space to account for the P_f of the fully stimulated bladder. The irony of all this, of course, is that the fusion of a long tubular vesicle with the mucosal membrane has functionally inserted into that membrane not only the water-permeable pores, but also along with them their own unstirred layer. The value of 17 for P_f/P_{d_w} thus becomes a rather sophisticated artifact of unstirred layers, and tells us nothing about the length of the narrow, single-file region in the water pores.

Comparison of the ADH-induced Pores to the Water Pores in the Red Cell Membrane. A striking phenomenological similarity exists between the ADH-induced pores and the red cell pores discussed in the preceeding chapter. Not only do both possess a narrow region that excludes nonelectrolytes as small as urea and through which single-file transport of water occurs, both also have a very low permeability to ions. As pointed out by Finkelstein and Rosenberg (1979), the conductance of the ADH-induced water pores is *at least* 10^4 times smaller than that of the gramicidin A pore. [Gluck and Al-Awqati (1980) measured their proton permeability and found it to be at least 200 times smaller than that of the gramicidin A pore.] This is not to say that the two types of pores are chemically identical. For one thing, the ADH-induced water pores are not inhibited by PCMBS (R. Hays and M. Rubin, personal communication). Nevertheless, it is interesting that nature apparently provides at least two examples of narrow pores that are permeable to water and virtually nothing else.

What can account for the low ion permeability of these biological water pores compared to that of the gramicidin A pore? The electrostatic energy required to move an ion from bulk solution into the center of a pore, and which enters as an exponential term in the calculation of ion permeability, is a sensitive function of the length of the pore, its radius, and the dielectric constant of the pore walls (e.g., Levitt, 1978a). If these water pores are twice the length of the gramicidin A pore (50 Å instead of 25 Å), of somewhat smaller radius, and/or have walls of lower dielectric constant, these factors could enormously reduce their conductance (ion permeability). Alternatively, the judicious placement of charge groups—a positive charge at one end to screen out cations and a negative charge at the other end to screen out anions—could also reduce their conductance. Ion permeability is such a sensitive function of electrostatic energy terms that it is no problem to envision pores of comparable dimensions with much different ion permeability characteristics. It is not necessary to assume a low dielectric constant region within the pore that interrupts the water continuum in order to account for its low ion (and particularly low H^+) permeability (Macey, 1984), although this is also a possibility.

Bibliography

Adamson, A.W. 1967. *Physical Chemistry of Surfaces,* 2nd Ed. Wiley-Interscience, New York, pp. 225–227.

Almers, W., and Stirling, C.E. 1984. The distribution of transport proteins over animal cell membranes. *J. Membrane Biol.* **77:** 169–186.

Al-Zahid, G., Schafer, J.A., Troutman, S.L., and Andreoli, T.E. 1977. Effect of antidiuretic hormone on water and solute permeation and the activation energies for these processes in mammalian cortical collecting tubules. *J. Membrane Biol.* **31:** 103–129.

Andersen, B., and Ussing, H.H. 1957. Solvent drag on non-electrolytes during osmotic flow through isolated toad skin and its response to antidiuretic hormone. *Acta Physiol. Scand.* **39:** 228–239.

Andersen, O.S. 1978. Ion transport across simple membranes. In *Renal Function,* Eds., G.H. Giebisch and E.F. Purcell. Josiah Macy Jr. Foundation, New York, pp. 71–99.

Andersen, O.S. 1983a. Ion movement through gramicidin A channels. Single channel measurements at very high potentials. *Biophys. J.* **41:** 119–133.

Andersen, O.S. 1983b. Ion movement through gramicidin A channels. Studies on the diffusion-controlled association step. *Biophys. J.* **41:** 147–165.

Andersen, O.S., and Procopio. J. 1980. Ion movement through gramicidin A channels. On the importance of the aqueous diffusion resistance and ion-water interactions. *Acta Physiol. Scand. Suppl.* **481:** 27–35.

Anderson, J.L., and Malone, D.M. 1974. Mechanism of osmotic flow in porous membranes. *Biophys. J.* **14:** 957–982.

Andreoli, T.E., and Monahan, M. 1968. The interaction of polyene antibiotics with thin lipid membranes. *J. Gen. Physiol.* **52:** 300–325.

Andreoli, T.E., Schafer, J.A., and Troutman, S.L. 1971. Coupling of solute and solvent flows in porous bilayer membranes. *J. Gen. Physiol.* **57:** 479–493.

Bamberg, E., Apell, H.-J., and Alpes, H. 1977. Structure of the gramicidin A channel: Discrimination between the $\Pi_{L,D}$ and the β helix by electrical measurements with lipid bilayer membranes. *Proc. Natl. Acad. Sci. USA* **74:** 2402–2406.

Barry, P.H., and Diamond, J. 1984. Effects of unstirred layers on membrane phenomena. *Physiol. Rev.* **64:** 763–872.

Bean, C.P. 1972. The physics of porous membranes—neutral pores. In *Membranes–A Series of Advances.* Vol. 3. *Macroscopic Systems and Models,* Ed., G. Eisenman. Marcel Dekker, New York, 1–54.

Bean, R.C., Shepherd, W.C., Chan H., and Eichner, J. 1969. Dicrete conductance fluctuations in lipid bilayer protein membranes. *J. Gen. Physiol.* **53:** 741–757.

Benga, G., Pop, V.I., Popescu, O., Ionescu, M., and Mihele, V. 1983. Water exchange through erythrocyte membranes: Nuclear magnetic resonance studies on the effects of inhibitors and of chemical modification of human membranes. *J. Membrane Biol.* **76:** 129–137.

Benz, R., Fröhlich, O., Läuger, P., and Montal, M. 1975. Electrical capacity of black lipid films and of lipid bilayers made from monolayers. *Biochim. Biophys. Acta* **394:** 323–334.

Benz, R., Tosteson, M.T., and Schubert, D. 1984. Formation and properties of tetramers of band 3 protein from human erythrocyte membranes in planar lipid bilayers. *Biochim. Biophys. Acta* **775:** 347–355.

Berry, C.A. 1983. Water permeability and pathways in the proximal tubule. *Amer. J. Physiol.* **245:** F279–F294.

Berry, C.A., and Boulpaep, E.L. 1975. Nonelectrolyte permeability of the paracellular pathway in *Necturus* proximal tubule. *Amer. J. Physiol.* **228:** 581–595.

Bikerman, J.J. 1958. *Surface Chemistry. Theory and Application,* 2nd Ed. Academic Press, New York, pp. 412–418.

Blok, M.C., Van Deenen, L.L.M., and de Gier, J. 1977. The effect of cholesterol incorporation on the temperature dependence of water permeation through liposomal membranes prepared from phosphatidylcholines. *Biochim. Biophys. Acta* **464:** 509–518.

Blum, R.M., and Forster, R.E. 1970. The water permeability of erythrocytes. *Biochim. Biophys. Acta* **203:** 410–423.

Boehler, B.A., de Gier, J., and Van Deenen, L.L.M. 1978. The effect of gramicidin A on the temperature dependence of water permeation through liposomal membranes prepared from phosphatidylcholines with different chain lengths. *Biochim. Biophys. Acta* **512:** 480–488.

BIBLIOGRAPHY

Brahm, J. 1977. Temperature-dependent changes of chloride transport kinetics in human red cells. *J. Gen. Physiol.* **70:** 283–306.

Brahm, J. 1982. Diffusional water permeability of human erythrocytes and their ghosts. *J. Gen. Physiol.* **79:** 791–819.

Brahm, J. 1983a. Permeability of human red cells to a homologous series of aliphatic alcohols. Limitations of the continuous flow-tube method. *J. Gen. Physiol.* **81:** 283–304.

Brahm, J. 1983b. Urea permeability of human red cells. *J. Gen. Physiol.* **82:** 1–23.

Brahm, J., and Wieth, J.O. 1977. Separate pathways for urea and water, and for chloride in chicken erythrocytes. *J. Physiol.* (London) **266:** 727–749.

Branton, D., Cohen, C.M., and Tyler, J. 1981. Interaction of cytoskeletal proteins on the human erythrocyte membrane. *Cell* **24:** 24–32.

Brown, M.F., Ribeiro, A.A., and Williams, G.D. 1983. New view of lipid bilayer dynamics from ^2H and ^{13}C NMR relaxation time measurements. *Proc. Natl. Acad. Sci. USA* **80:** 4325–4329.

Brown, P.A., Feinstein, M.D., and Sha'afi, R.I. 1975. Membrane proteins related to water transport in human erythrocytes. *Nature* **254:** 523–525.

Brutyan, R.A., and Yermishkin, L.N. 1983. Interaction of ions in amphotericin channels. *Biophysics* **28:** 465–469. [Translated from *Biofizika* **28:** 436–439, 1983.]

Cabantchik, Z.I., Knauf, P.A., and Rothstein, A. 1978. The anion transport system of the red blood cell. The role of membrane protein evaluated by the use of 'probes'. *Biochim. Biophys. Acta* **515:** 239–302.

Carruthers, A., and Melchior, D.L. 1983. Studies on the relationship between bilayer water permeability and bilayer physical state. *Biochemistry* **22:** 5797–5807.

Carvounis, C.P., Franki, N., Levine, S.D., and Hays, R.M. 1979. Membrane pathways for water and solutes in toad bladder: I. Independent activation of water and urea transport. *J. Membrane Biol.* **49:** 253–268.

Cass, A., and Dalmark, M. 1973. Equilibrium dialysis of ions in nystatin-treated red cells. *Nature* **244:** 47–49.

Cass, A., and Finkelstein, A. 1967. Water permeability of thin lipid membranes. *J. Gen. Physiol.* **50:** 1765–1784.

Cass, A., Finkelstein, A., and Krespi, V. 1970. The ion permeability induced in thin lipid membranes by the polyene antibiotics nystatin and amphotericin B. *J. Gen. Physiol.* **56:** 100–124.

BIBLIOGRAPHY

Collander, R., and Bärlund, H. 1933. Permeabilitatsstudien an *Chara ceratophylla. Acta Bot. Fenn.* **11**: 1–114.

Conti, F., De Felice, L.J., and Wanke, E. 1975. Potassium and sodium ion current noise in the membrane of the squid giant axon. *J. Physiol.* (London) **248**: 45–82.

Conti, F., and Neher, E. 1980 . Single channel recordings of K^+ currents in squid axons. *Nature* **285**: 140–143.

Dainty, J. 1963. Water relations of plant cells. *Adv. Botan. Res.* **1**: 279–326.

Dainty, J., and House, C.R. 1966. An examination of the evidence for membrane pores in frogskin. *J. Physiol.* (London) **185**: 172–184.

Dani, J.A., and Levitt, D.G. 1981a. Binding constants for Li^+, K^+ and Tl^+ in the gramicidin channel determined from water permeability measurements. *Biophys. J.* **35**: 485–500.

Dani, J.A., and Levitt, D.G. 1981b. Water transport and ion-water interaction in the gramicidin channel. *Biophys. J.* **35**: 501–508.

Dawson, D.C. 1982. Thermodynamic aspects of radiotracer flow. In *Biological Transport of Radiotracers,* Ed., L.G. Colombetti. CRC Press, Boca Raton, Florida, pp. 79–95.

Dennis, V.W., Stead, N.W., and Andreoli, T.E. 1970. Molecular aspects of polyene- and sterol-dependent pore formation in thin lipid membranes. *J. Gen. Physiol.* **55**: 375–400.

Diamond, J.M. 1962. The mechanism of water transport by the gall-bladder. *J. Physiol.* (London) **161**: 503–527.

Diamond, J.M. 1979. Osmotic water flow in leaky epithelia. *J. Membrane Biol.* **51**: 195–216.

Di Bona, D.R., Civan, M.M., and Leaf, A., 1969. The cellular specificity of the effect of vasopressin on toad urinary bladder. *J. Membrane Biol.* **1**: 79–91.

Dick, D.A.T. 1966. *Cell Water.* Butterworths, Washington, D.C., pp. 108–111.

Dix, J.A., and Solomon, A.K. 1984. Role of membrane proteins and lipids in water diffusion across red cell membranes. *Biochim. Biophys. Acta* **773**: 219–230.

Dubos, R.J. 1939. Studies on a bactericidal agent extracted from a soil bacillus. I. Preparation of the agent. Its activity *in vitro. J. Exp. Med.* **70**: 1–10.

Dunham, P.B., Cass, A., Trinkaus, J.P., and Bennett, M.V.L. 1970. Water permeability of *Fundulus* eggs. *Biol. Bull.* **139**: 420–421.

Durbin, R.P. 1960. Osmotic flow of water across permeable cellulose membranes. *J. Gen. Physiol.* **44:** 315–326.

Durbin, R.P., Frank, H., and Solomon, A.K. 1956. Water flow through frog gastric mucosa. *J. Gen. Physiol.* **39:** 535–551.

Eggena, P. 1972. Temperature dependence of vasopressin action on the toad bladder. *J. Gen. Physiol.* **59:** 519–533.

Ehrenstein, G., Lecar, H., and Nossal, R. 1970. The nature of the negative resistance in bimolecular lipid membranes containing excitability-inducing material. *J. Gen. Physiol.* **55:** 119–133.

Einstein, A. 1905. On the movement of small particles suspended in a stationary liquid demanded by the molecular-kinetic theory of heat. (Translated from *Ann. d. Phys.* **17:** 549–560.) In *Investigations on the Theory of the Brownian Movement* by Albert Einstein, edited with notes by R. Fürth, translated by A.D. Cowper. Dover, New York, 1956, pp. 1–18.

Einstein, A. 1908. The elementary theory of the Brownian motion. (Translated from *Zeit. Für Elektrochemie* **14:** 235–239, 1908.) In *Investigations on the Theory of the Brownian Movement* by Albert Einstein, edited with notes by R. Fürth, translated by A.D. Cowper. Dover, New York, 1956, pp. 68–85.

Ermishkin, L.N., Kasumov, Kh. M., and Potzeluyev, V.M. 1976. Single ionic channels induced in lipid bilayers by polyene antibiotics amphotericin B and nystatine. *Nature* **262:** 698–699.

Ermishkin, L.N., Kasumov, Kh. M., and Potzeluyev, V.M. 1977. Properties of amphotericin B channels in a lipid bilayer. *Biochim. Biophys. Acta* **470:** 357–367.

Etchebest, C., and Pullman, A. 1984. The gramicidin A channel. Role of the ethanolamine end chain on the energy profile for single occupancy by Na^+. *FEBS Letters* **170:** 191–195.

Everitt, C.T., and Haydon, D.A. 1969. Influence of diffusion layers during osmotic flow across bimolecular lipid membranes. *J. Theoret. Biol.* **22:** 9–19.

Everitt, C.T., Redwood, W.R., and Haydon, D.A. 1969. Problem of boundary layers in exchange diffusion of water across bimolecular lipid membranes. *J. Theoret. Biol.* **22:** 20–32.

Fabry, M.E., and Eisenstadt, M. 1975. Water exchange between red cells and plasma. Measurements by nuclear magnetic relaxation. *Biophys. J.* **15:** 1101–1110.

BIBLIOGRAPHY

Farmer, R.E.L., and Macey, R.I. 1970. Perturbation of red cell volume: Rectification of osmotic flow. *Biochim. Biophys. Acta* **196**: 53–65.

Ferguson, D.R., and Twite, B.R. 1974. Effects of vasopressin on toad bladder membrane proteins: Relationship to transport of sodium and water. *J. Endocrinol.* **61**: 501–507.

Ferry, J.D. 1936. Statistical evaluation of sieve constants in ultrafiltration. *J. Gen. Physiol.* **20**: 95–104.

Fettiplace, R. 1978. The influence of the lipid on the water permeability of artificial membranes. *Biochim. Biophys. Acta* **513**: 1–10.

Fettiplace, R., Andrews, D.M., and Haydon, D.A. 1971. The thickness, composition and structure of some lipid bilayers and natural membranes. *J. Membrane Biol.* **5**: 277–296.

Fettiplace, R., and Haydon, D.A. 1980. Water permeability of lipid membranes. *Physiol. Rev.* **60**: 510–550.

Finer-Moore, J., and Stroud, R.M. 1984. Amphipathic analysis and possible formation of the ion channel in an acetylcholine receptor. *Proc. Natl. Acad. Sci. USA* **81**: 155–159.

Finkelstein, A. 1974. Aqueous pores created in thin lipid membranes by the antibiotics nystatin, amphotericin B and gramicidin A: Implications for pores in plasma membranes. In *Drugs and Transport Processes,* Ed., B. A. Callingham. MacMillan, London, pp. 241–250.

Finkelstein, A. 1976a. Water and nonelectrolyte permeability of lipid bilayer membranes. *J. Gen. Physiol.* **68**: 127–135.

Finkelstein, A. 1976b. Nature of the water permeability increase induced by antidiuretic hormone (ADH) in toad urinary bladder and related tissues. *J. Gen. Physiol.* **68**: 137–143.

Finkelstein, A., and Andersen, O.S. 1981. The gramicidin A channel: A review of its permeability characteristics with special reference to the single-file aspects of transport. *J. Membrane Biol.* **59**: 155–171.

Finkelstein, A., and Cass, A. 1968. Permeability and electrical properties of thin lipid membranes. *J. Gen. Physiol.* **52**: 145s–172s.

Finkelstein, A., and Holz, R. 1973. Aqueous pores created in thin lipid membranes by the polyene antibiotics nystatin and amphotericin B. In *Membranes, A Series of Advances.* Vol. 2. *Lipid Bilayers and Antibiotics,* Ed., G. Eisenman. Marcel Dekker, New York, pp. 377–408.

Finkelstein, A., and Rosenberg, P.A. 1979. Single-file transport: Implications for ion and water movement through gramicidin A channels. In

Membrane Transport Processes, Vol. 3, Eds., C.F. Stevens and R.W. Tsien. Raven Press, New York, pp. 73–88.

Fischbarg, J. 1982. The paracellular pathway in the corneal endothelium. In *The Paracellular Pathway,* Eds., S.E. Bradley and E.F. Purcell. Josiah Macy, Jr. Foundation, New York, pp. 307–318.

Fornili, S.L., Vercauteren, D.P., and Clementi, F. 1984. Water structure in the gramicidin A transmembrane channel. *J. Mol. Catalysis* **23**: 341–356.

Frömter, E. 1972. The route of passive ion movement through the epithelium of *Necturus* gallbladder. *J. Membrane Biol.* **8**: 259–301.

Frömter, E., and Diamond, J.M. 1972. Route of passive ion permeability in epithelia. *Nature (New Biol.)* **235**: 9–13.

Garby, L., 1957. Studies of the transfer of matter across membranes with special reference to the isolated human amniotic membrane and the exchange of amniotic fluid. *Acta Physiol. Scand.* **40** (Suppl. 137): 1–84.

Garrett, H.E. 1961. Aggregation in detergent solutions. In *Surface Activity and Detergency,* Ed., K. Durham. MacMillan, London, pp. 29–51.

Ginzburg, B.Z., and Katchalsky, A. 1963. The frictional coefficients of the flows of non-electrolytes through artificial membranes. *J. Gen. Physiol.* **47**: 403–418.

Glickson, J.D., Mayers, D.F., Settine, J.M., and Urry, D.W. 1972. Spectroscopic studies on the conformation of gramicidin A. Proton magnetic resonance assignments, coupling constants, and H-D exchange. *Biochemistry* **11**: 477–486.

Gluck, S., and Al-Awqati, Q. 1980. Vasopressin increases water permeability by inducing pores. *Nature* **284**: 631–632.

Goldstein, D.A., and Solomon, A.K. 1960. Determination of equivalent pore radius for human red cells by osmotic pressure measurement. *J. Gen. Physiol.* **44**: 1–17.

Gonzalez, E., Carpi-Medina, P., and Whittenbury, G. 1982. Cell osmotic water permeability of isolated rabbit proximal straight tubules. *Amer. J. Physiol.* **242**: F321–F330.

Gorter, E., and Grendel, F. 1925. On bimolecular layers of lipoids on the chromocytes of the blood. *J. Exp. Med.* **41**: 439–443.

Granthum, J.J., and Burg, M.B. 1966. Effect of vasopressin and cyclic AMP on permeability of isolated collecting tubules. *Am. J. Physiol.* **211**: 255–259.

BIBLIOGRAPHY

Graziani, Y., and Livne, A. 1972. Water permeability of bilayer lipid membranes: Sterol-lipid interaction. *J. Membrane Biol.* **7**: 275–284.

de Groot, S.R. 1958. *Thermodynamics of Irreversible Processes.* North Holland Publishing Company, Amsterdam, pp 1–242.

Gutknecht, J. 1967. Membranes of *Valonia ventricosa:* Apparent absence of water-filled pores. *Science* **158**: 787–788.

Guy, H.R. 1984. A structural model of the acetylcholine receptor channel based on partition energy and helix packing calculations. *Biophys. J.* **45**: 249–261.

Hammond, S.M. 1977. Biological activity of polyene antibiotics. In *Progress in Medicinal Chemistry,* Vol. 14, Eds., G.P. Ellis and G.B. West. Elsevier North-Holland Biomedical Press, Amsterdam, pp. 106–179.

Hanai, T., and Haydon, D.A. 1966. The permeability to water of biomolecular lipid membranes. *J. Theoret. Biol.* **11**: 370–382.

Hanai, T., Haydon, D.A., and Taylor, J. 1964. An investigation by electrical methods of lecithin-in-hydrocarbon films in aqueous solutions. *Proc. Roy. Soc.* (London), Ser. A. **281**: 377–391.

Handler, J.S., and Orloff, J. 1973. The mechanism of action of antidiuretic hormone. In *Handbook of Physiology: Renal Physiology,* Eds., S.R. Geiger, J. Orloff, and R.W. Berliner. Williams & Wilkins, Baltimore, pp. 791–814.

Hartley, G.S., and Crank, J. 1949. Some fundamental definitions and concepts in diffusion processes. *Trans. Faraday Soc.* **45**: 801–818.

Haydon, D.A., Hendry, B.M., Levinson, S.R., and Requena, J. 1977. Anesthesia by the n-alkanes. A comparative study of nerve impulse blockage and the properties of black lipid bilayer membranes. *Biochim. Biophys. Acta* **470**: 17–34.

Hays, R.M. 1972. The movement of water across vasopressin-sensitive epithelia. *Current Topics in Membranes and Transport* **3**: 339–366.

Hays, R.M., and Franki, N., 1970. The role of water diffusion in the action of vasopressin. *J. Membrane Biol.* **2**: 263–276.

Hays, R.M., and Leaf, A. 1962. Studies on the movement of water through the isolated toad bladder and its modification by vasopressin. *J. Gen. Physiol.* **45**: 905–919.

Hayward, A.T.J. 1971. Negative pressure in liquids: Can it be harnessed to serve man? *Amer. Sci.* **59**: 434–443.

Hebert, S.C., and Andreoli, T.E. 1980. Interactions of temperature and

ADH on transport processes in cortical collecting tubules. *Am. J. Physiol.* **238:** F470–F480.

Hebert, S.C., Culpepper, R.M., and Andreoli, T.E. 1981. NaCl transport in mouse medullary thick ascending limbs. I. Functional nephron heterogeneity and ADH-stimulated NaCl cotransport. *Amer. J. Physiol.* **241:** F412–F431.

Henderson, R., and Unwin, P.N.T. 1975. Three-dimensional model of purple membrane obtained by electron microscopy. *Nature* **257:** 28–32.

Hevesy, G., Hofer, E., and Krogh, A. 1935. The permeability of the skin of frogs to water as determined by D_2O and H_2O. *Skandinav. Archiv. Physiol.* **72:** 199–214.

Hill, A. 1982. Osmosis: A bimodal theory with implications for symmetry. *Proc. Royal Soc. Lond. B.* **215:** 155–174.

Hille, B. 1975. Ionic selectivity of Na and K channels of nerve membranes. In *Membranes: A Series of Advances, Vol. 3. Lipid Bilayers and Biological Membranes: Dynamic Properties,* Ed., G. Eisenman, Marcel Dekker, New York, pp. 255–323.

Hille, B. 1984. *Ionic Channels of Excitable Membranes.* Sinauer Associates, Sunderland, Mass., pp. 205–225.

Hladky, S.B., and Haydon, D.A. 1970. Discreteness of conductance change in bimolecular lipid membranes in the presence of certain antibiotics. *Nature* **225:** 451–453.

Hladky, S.B., and Haydon, D.A. 1972. Ion transfer across lipid membranes in the presence of gramicidin A. I. Studies on the unit conductance channel. *Biochim. Biophys. Acta* **274:** 294–312.

Hladky, S.B., and Haydon, D.A. 1984. Ion movement in gramicidin channels. *Current Topics in Membranes and Transport* **21:** 327–372.

Hodgkin, A.L., Huxley, A.F., and Katz, B. 1952. Measurement of current-voltage relations in the membrane of the giant axon of *Loligo. J. Physiol.* (London) **116:** 424–448.

Hodgkin, A.L., and Keynes, R.D. 1955. Active transport of cations in giant axons from *Sepia* and *Loligo. J. Physiol.* (London) **128:** 28–60.

Holz, R., and Finkelstein A. 1970. The water and nonelectrolyte permeability induced in thin lipid membranes by the polyene antibiotics nystatin and amphotericin B. *J. Gen. Physiol.* **56:** 125–145.

Hotchkiss, R.D., and Dubos, R.J. 1940. Fractionation of the bactericidal agent from cultures of a soil bacillus. *J. Biol. Chem.* **132:** 791–792.

Huang, C., and Thompson, T.E. 1966. Properties of lipid bilayer membranes separating two aqueous phases: Water permeability. *J. Mol. Biol.* **15**: 539–554.

Hunter, M.J. 1977. Human erythrocyte anion permeabilities measured under conditions of net charge transfer. *J. Physiol.* (London) **268**: 35–49.

Jay, A.W.L. 1975. Geometry of the human erythrocyte. I. Effect of albumin on cell geometry. *Biophys. J.* **15**: 205–222.

Jay, D.G. 1983. Characterization of the chicken erythrocyte anion exchange protein. *J. Biol. Chem.* **258**: 9431–9436.

Kachadorian, W.A., Levine, S.D., Wade, J.B., Di Scala, V.A., and Hays, R.M. 1977a. Relationship of aggregated intramembranous particles to water permeability in vasopressin-treated toad urinary bladder. *J. Clin. Invest.* **59**: 576–581.

Kachadorian, W.A., Muller, J., and Finkelstein, A. 1981. Role of osmotic forces in exocytosis: Studies of ADH-induced fusion in toad urinary bladder. *J. Cell Biol.* **91**: 584–588.

Kachadorian, W.A., Muller, J., Rudich, S.W., and Di Scala, V.A. 1979. Temperature dependence of ADH-induced water flow and intramembranous aggregates in toad bladder. *Science* **205**: 910–913.

Kachadorian, W.A., Wade, J.B., and Di Scala, V.A. 1975. Vasopressin: Induced structural change in toad bladder luminal membrane. *Science* **190**: 67–69.

Kachadorian, W.A., Wade, J.B., Uiterwyk, C.C. and Di Scala, V.A. 1977b. Membrane structure and functional responses to vasopressin in toad bladder. *J. Membrane Biol.* **30**: 381–401.

Karan, D.M., and Macey, R.I. 1980. The permeability of the human red cell to deuterium oxide (heavy water). *J. Cell. Physiol.* **104**: 209–214.

Karlin, A. 1983. The anatomy of a receptor. *Neuroscience Commentaries* **1**: 111–123.

Katchalsky, A., and Curran, P.F. 1965. *Nonequilibrium Thermodynamics in Biophysics.* Harvard University Press, Cambridge, Mass., pp 1–248.

Kedem, O., and Katchalsky, A. 1958. Thermodynamic analysis of the permeability of biological membranes to non-electrolytes. *Biochim. Biophys. Acta* **27**: 229–246.

Kedem, O., and Katchalsky, A. 1961. A physical interpretation of the phenomenological coefficients of membrane permeability. *J. Gen. Physiol.* **45**: 143–179.

BIBLIOGRAPHY

Kleinberg, M.E., and Finkelstein, A. 1984. Single-length and double-length channels formed by nystatin in lipid bilayer membranes. *J. Membrane Biol.* **80**: 257–269.

Knauf, P.A. 1979. Erythrocyte anion exchange and the band 3 protein: Transport kinetics and molecular structure. *Current Topics in Membranes and Transport* **12**: 249–363.

Knauf, P.A., Law, F.-Y., and Marchant, P.J. 1983. Relationship of net chloride flow across the human erythrocyte membrane to the anion exchange mechanism. *J. Gen. Physiol.* **81**: 95–126.

Knauf, P.A., Proverbio, F., and Hoffman, J.F. 1974. Chemical characterization and pronase susceptibility of the Na:K pump-associated phosphoprotein of human red blood cells. *J. Gen. Physiol.* **63**: 305–323.

Knauf, P.A., and Rothstein, A. 1971. Chemical modification of membranes. I. Effects of sulfhydral and amino reactive reagents on anion and cation permeability of the human red blood cell. *J. Gen. Physiol.* **58**: 190–210.

Koefoed-Johnsen, V., and Ussing, H.H. 1953. The contribution of diffusion and flow to the passage of D_2O through living membranes. Effect of neurohypophyseal hormone on isolated anuran skin. *Acta Physiol. Scand.* **28**: 60–76.

Kohler, H.-H., and Heckmann, K. 1979. Unidirectional fluxes in saturated single-file pores of biological and artificial membranes. I. Pores containing no more than one vacancy. *J. Theoret. Biol.* **79**: 381–401.

Kohler, H.-H., and Heckmann, K. 1980. Unidirectional fluxes in saturated single-file pores of biological and artificial membranes. II. Asymptotic behavior at high degrees of saturation. *J. Theoret. Biol.* **85**: 575–595.

Kopito, R.R., and Lodish, H.F. 1985. Primary structure and transmembrane orientation of the murine anion exchange protein. *Nature* **316**: 234–238.

de Kruijff, B., and Demel, R.A. 1974. Polyene antibiotic-sterol interactions in membrane of *Acholeplasma laidlawii* cells and lecithin liposomes: III. Molecular structure of the polyene antibiotic cholesterol complexes. *Biochim. Biophys. Acta* **339**: 57–70.

de Kruijff, B., Gerritsen, W.J., Oerlemans, A., Demel, R.A., and van Deenen, L.L.M. 1974. Polyene antibiotic-sterol interactions in membranes of *Acholeplasma laidlawii* cells and lecithin liposomes. I. Specificity of the membrane permeability changes induced by the polyene antibiotics. *Biochim. Biophys. Acta* **339**: 30–43.

Laris, P.C. 1958. Permeability and utilization of glucose in mammalian erythrocytes. *J. Cell. Comp. Physiol.* **51**: 273–307.

Lawaczek, R. 1979. On the permeability of water molecules across vesicular lipid bilayers. *J. Membrane Biol.* **51**: 229–261.

Lea, E.J.A. 1963. Permeation through long narrow pores. *J. Theoret. Biol.* **5**: 102–107.

Leaf, A., and Hays, R.M. 1962. Permeability of the isolated toad bladder to solutes and its modification by vasopressin. *J. Gen. Physiol.* **45**: 921–932.

Lecuyer, H., and Dervichian, D.G. 1969. Structure of aqueous mixtures of lecithin and cholesterol. *J. Mol. Biol.* **45**: 39–57.

LeFevre, P.G. 1954. The evidence for active transport of monosaccharides across the red cell membrane. In *Symposium of the Society for Experimental Biology. No. 8. Active Transport and Secretion.* Academic Press, New York, pp. 118–135.

Levine, S., Franki, N., and Hays, R.M. 1973. Effect of phloretin on water and solute movement in the toad bladder. *J. Clin. Invest.* **52**: 1435–1442.

Levine, S.D., Jacoby, M., and Finkelstein, A. 1984a. The water permeability of toad urinary bladder. I. Permeability of barriers in series with the luminal membrane. *J. Gen. Physiol.* **83**: 529–541.

Levine, S.D., Jacoby, M., and Finkelstein, A. 1984b. The water permeability of toad urinary bladder. II. The value of $P_f/P_d(w)$ for the antidiuretic hormone-induced water permeation pathway. *J. Gen. Physiol.* **83**: 543–561.

Levine, S.D., and Kachadorian, W.A. 1981. Barriers to water flow in vasopressin-treated toad urinary bladder. *J. Membrane Biol.* **61**: 135–139.

Levine, S.D., Levine, R.D., Worthington, R.E., and Hays, R.M. 1976. Selective inhibition of osmotic water flow by general anesthetics in toad urinary bladder. *J. Clin. Invest.* **58**: 980–988.

Levitt, D.G. 1973. Kinetics of diffusion and convection in 3.2-Å pores. Exact solution by computer simulation. *Biophys. J.* **13**: 186–206.

Levitt, D.G. 1974. A new theory of transport for cell membrane pores. I. General theory and application to red cell. *Biochim. Biophys. Acta* **373**: 115–131.

Levitt, D.G. 1975. General continuum analysis of transport through pores. I. Proof of Onsager's reciprocity postulate for uniform pore. *Biophys. J.* **15**: 533–551.

Levitt, D.G. 1978a. Electrostatic calculations for an ion channel. I. Energy and potential profiles and interactions between ions. *Biophys. J.* **22**: 209–219.

Levitt, D.G. 1978b. Electrostatic calculations for an ion channel. II. Kinetic behavior of the gramicidin A channel. *Biophys. J.* **22**: 221–248.

Levitt, D.G. 1984. Kinetics of movement in narrow channels. *Current Topics in Membranes and Transport* **21**: 181–198.

Levitt, D.G., Elias, S.R., and Hautman, J.M. 1978. Number of water molecules coupled to the transport of sodium, potassium and hydrogen ions via gramicidin, nonactin or valinomycin. *Biochim. Biophys. Acta* **512**: 436–451.

Levitt, D.G., and Mlekoday, H.J. 1983. Reflection coefficient and permeability of urea and ethylene gycol [sic] in the human red cell membrane. *J. Gen. Physiol.* **81**: 239–253.

Levitt, D.G., and Subramanian, G. 1974. A new theory of transport for cell membrane pores. II. Exact results and computer simulation (molecular dynamics). *Biochim. Biophys. Acta* **373**: 132–140.

Lichtenstein, N.S., and Leaf, A. 1965. Effect of amphotericin B on the permeability of the toad bladder. *J. Clin. Invest.* **44**: 1328–1342.

Lieb, W.R., and Stein, W.D. 1971. The molecular basis of simple diffusion within biological membranes. *Current Topics in Membranes and Transport* **2**: 1–39.

Lin, S., and Snyder, C.E. 1977. High affinity cytochalasin B binding to red cell membrane proteins which are unrelated to sugar transport. *J. Biol. Chem.* **252**: 5464–5471.

Longuet-Higgins, H.C., and Austin, G. 1966. The kinetics of osmotic transport through pores of molecular dimensions. *Biophys. J.* **6**: 217–224.

Lorenz, P.B. 1952. The phenomenology of electro-osmosis and streaming potential. *J. Phys. Chem.* **56**: 775–778.

Luzzati, V., and Husson, F. 1962. The structure of the liquid-crystalline phases of lipid-water systems. *J. Cell Biol.* **12**: 207–219.

MacDonald, A.G. 1972. The role of high hydrostatic pressure in the physiology of marine animals. In *Symposium of the Society for Experimental Biology,* No. 26: *The Effects of Pressure on Organisms,* Eds., M.A. Sleigh and A.G. MacDonald. Academic Press, New York, pp. 209–231.

Macey, R.I. 1979. Transport of water and nonelectrolytes across red cell membranes. In *Membrane Transport in Biology,* Vol. 2: *Transport*

BIBLIOGRAPHY

Across Single Biological Membranes, Eds., G. Giebisch, D.C. Tosteson, and H.H. Ussing. Springer-Verlag, New York, pp. 1–57.

Macey, R.I. 1984. Transport of water and urea in red blood cells. *Amer. J. Physiol.* **246:** C195–C203.

Macey, R.I., and Farmer, R.E.L. 1970. Inhibition of water and solute permeability in human red cells. *Biochim. Biophys. Acta* **211:** 104–106.

Macey, R.I., Karan, D.M., and Farmer, R.E.L. 1972. Properties of water channels in human red cells. In *Biomembranes,* Vol. 3: *Passive Permeability of Cell Membranes,* Eds., F. Kreuzer and J.F.G. Slegers. Plenum Press, New York, pp. 331–340.

MacKay, D.H.J., Berens, P.H., Wilson, K.R., and Hagler, A.T. 1984. Structure and dynamics of ion transport through gramicidin A. *Biophys. J.* **46:** 229–248.

Manning, G.S. 1975. The relation between osmotic flow and tracer solvent diffusion for single-file transport. *Biophysical Chem.* **3:** 147–152.

Manning, G.S. 1976. Deviation from the Einstein relation of the single-file diffusion coefficient. *Biophys. Chem.* **5:** 389–394.

Marchesi, V.T. 1979. Functional proteins of the human red blood cell membrane. *Semin. Hematol.* **16:** 3–20.

Martin, J.F. 1977. Biosynthesis of polyene macrolide antibiotics. *Ann. Rev. Microbiol.* **31:** 13–38.

Marty, A., and Finkelstein, A. 1975. Pores formed in lipid bilayer membranes by nystatin. Differences in its one-sided and two-sided action. *J. Gen. Physiol.* **65:** 515–526.

Mauro, A. 1957. Nature of solvent transfer in osmosis. *Science* **126:** 252–253.

Mauro, A. 1960. Some properties of ionic and nonionic semipermeable membranes. *Circulation* **21:** 845–854.

Mauro, A. 1965. Osmotic flow in a rigid porous membrane. *Science* **149:** 867–869.

Mauro, A. 1981. The role of negative pressure in osmotic equilibrium and osmotic flow. In *Alfred Benzon Symposium 15: Water Transport Across Epithelia,* Eds., H.H. Ussing, N. Bindslev, N.A. Lassen, and O. Sten-Knudsen. Munksgaard, Copenhagen, pp. 107–119.

Mayrand, R.R., and Levitt, D.G. 1983. Urea and ethylene glycol-facilitated transport systems in the human red cell membrane. Saturation, competition, and asymmetry. *J. Gen. Physiol.* **81:** 221–237.

Medoff, G., and Kobayashi, G.A. 1980. The polyenes. In *Antifungal Chemotherapy,* Ed., D.C.E. Speller. John Wiley & Sons, New York, pp. 3–33.

Meschia, G., and Setnikar, I. 1958. Experimental study of osmosis through a collodian membrane. *J. Gen. Physiol.* **42:** 429–444.

Miller, C. 1982. Feeling around inside a channel in the dark. In *Transport in Biomembranes: Model Systems and Reconstitution,* Eds., R. Antolini, A. Gliozzi, and A. Gorio. Raven Press, New York, pp. 99–108.

Mlekoday, H.J., Moore, R., and Levitt, D.G. 1983. Osmotic water permeability of the human red cell. Dependence on direction of water flow and cell volume. *J. Gen. Physiol.* **81:** 213–220.

Montal, M., and Mueller, P. 1972. Formation of bimolecular membranes from lipid monolayers and a study of their electrical properties. *Proc. Natl. Acad. Sci. USA* **69:** 3561–3566.

Moura, T.F., Macey, R.I., Chien, D.Y., Karan, D., and Santos, H. 1984. Thermodynamics of all-or-none water channel closure in red cells. *J. Membrane Biol.* **81:** 105–111.

Mueller, P., Rudin, D.O., Tien, H.T., and Wescott, W.C. 1962. Reconstitution of excitable cell membrane structure *in vitro. Circulation* **26:** 1167–1171.

Mueller, P., Rudin, D.O., Tien, H.T., and Wescott, W.C. 1963. Methods for the formation of single bimolecular lipid membranes in aqueous solution. *J. Phys. Chem.* **67:** 534–535.

Muller, J., Kachadorian, W.A., and Di Scala, V.A. 1980. Evidence that ADH-stimulated intramembrane particle aggregates are transferred from cytoplasmic to luminal membranes in toad bladder epithelial cells. *J. Cell. Biol.* **85:** 83–95.

Myers, V.B., and Haydon, D.A. 1972. Ion transfer across lipid membranes in the presence of gramicidin A. II. The ion selectivity. *Biochim. Biophys. Acta* **274:** 313–322.

Naccache, P., and Sha'afi, R.I. 1974. Effect of PCMBS on water transfer across biological membranes. *J. Cell. Physiol.* **83:** 449–456.

Neher, E., Sandblom, J., and Eisenman, G. 1978. Ionic selectivity, saturation and block in gramicidin A channels. II. Saturation behavior of single channel conductances and evidence for the existence of multiple binding sites in the channel. *J. Membrane Biol.* **40:** 97–116.

Nevis, A.H. 1958. Water transport in invertebrate peripheral nerve. *J. Gen. Physiol.* **41:** 927–958.

Orbach, E., and Finkelstein, A. 1980. The nonelectrolyte permeability of planar lipid bilayer membranes. *J. Gen. Physiol.* **75**: 427–436.

Overton, E. 1895. Ueber die osmotischen Eigenschaften der lebenden Pflanzen und Tierzelle. *Vieteljahresschr. Naturforsch. Ges. Zurich* **40**: 159–201.

Owen, J.D., and Eyring, E.M. 1975. Reflection coefficients of permeant molecules in human red cell suspensions. *J. Gen. Physiol.* **66**: 251–265.

Paganelli, C.V., and Solomon, A.K. 1957. The rate of exchange of tritiated water across the human red cell membrane. *J. Gen. Physiol.* **41**: 259–277.

Pappenheimer, J.R. 1953. Passage of molecules through capillary walls. *Physiol. Rev.* **33**: 387–423.

Parisi, M., and Bourguet, J. 1983. The single-file hypothesis and water channels induced by antidiuretic hormone. *J. Membrane Biol.* **71**: 189–193.

Paulus, H., Sarkar, N., Mukherjee, P.K., Langley, D., Ivanov, V.T., Shepel, E.N., and Veatch, W. 1979. Comparison of the effect of linear gramicidin A analogues on bacterial sporulation, membrane permeability, and ribonucleic acid polymerase. *Biochemistry* **18**: 4532–4536.

Persson, B.-E., and Spring, K.R. 1982. Gallbladder epithelial cell hydraulic water permeability and volume regulation. *J. Gen. Physiol.* **79**: 481–505.

Pfeffer, W. 1887. *Osmotische Untersuchungen.* Engelmann, Leipzig. Now available in English translation as: *Osmotic Investigations* by W. Pfeffer. Translated by G.R. Kepner and E.J. Stadelman. Van Nostrand Reinhold, New York, 1985.

Pietras, R.J., and Wright, E.M. 1975. The membrane action of antidiuretic hormone (ADH) on toad urinary bladder. *J. Membrane Biol.* **22**: 107–123.

Pinto da Silva, P. 1973. Membrane intercalated particles in human erythrocyte ghosts: Sites of preferred passage of water molecules at low temperature. *Proc. Natl. Acad. Sci. USA* **70**: 1339–1343.

Pinto da Silva, P., and Nicolson, G.L. 1974. Freeze-etch localization of concanavalin A receptors to the membrane intercalated particles of human erythrocyte ghost membranes. *Biochim. Biophys. Acta* **363**: 311–319.

Poznansky, M., Tong, S., White, P.C., Milgram, J.M., and Solomon, A.K. 1976. Nonelectrolyte diffusion across lipid bilayer systems. *J. Gen. Physiol.* **67**: 45–66.

Prescott, D.M., and Zeuthen, E. 1953. Comparison of water diffusion and water filtration across cell surfaces. *Acta Physiol. Scand.* **28**: 77–94.

Procopio, J., and Andersen, O.S. 1979. Ion tracer fluxes through gramicidin A modified lipid bilayers. *Biophys. J.* **25**: 8a.

Pullman, A., and Etchebest, C. 1983. The gramicidin A channel: The energy profile for single and double occupancy in a head-to-head $\beta_{3,3}^{6,3}$-helical dimer backbone. *FEBS Letters* **163**: 199–202.

Redwood, W.R., and Haydon, D.A. 1969. Influence of temperature and membrane composition on the water permeability of lipid bilayers. *J. Theoret. Biol.* **22**: 1–8.

Redwood, W.R., Rall, E., and Perl, W. 1974. Red cell membrane permeability deduced from bulk diffusion coefficients. *J. Gen. Physiol.* **64**: 706–729.

Reeves, J.P., and Dowben, R.M. 1970. Water permeability of phospholipid vesicles. *J. Membrane Biol.* **3**: 123–141.

Renkin, E.M. 1954. Filtration, diffusion, and molecular sieving through porous cellulose membranes. *J. Gen. Physiol.* **38**: 225–243.

Rice, S.A. 1980. Hydrodynamic and diffusion considerations of rapid mix experiments with red blood cells. *Biophys. J.* **29**: 65–77.

Rich, G.T., Sha'afi, R.I., Romualdez, A., and Solomon, A.K. 1968. Effect of osmolality on the hydraulic permeability coefficient of red cells. *J. Gen. Physiol.* **52**: 941–954.

Rosenberg, P.A., and Finkelstein, A. 1978a. Interaction of ions and water in gramicidin A channels. Streaming potentials across lipid bilayer membranes. *J. Gen. Physiol.* **72**: 327–340.

Rosenberg, P.A., and Finkelstein, A. 1978b. Water permeability of gramicidin A-treated lipid bilayer membranes. *J. Gen. Physiol.* **72**: 341–350.

Sarges, R., and Witkop, B. 1965a. Gramicidin A. V. The structure of valine- and isoleucine-gramicidin A. *J. Amer. Chem. Soc.* **87**: 2011–2020.

Sarges, R., and Witkop, B. 1965b. Gramicidin A. VI. The synthesis of valine- and isoleucine-gramicidin A. *J. Amer. Chem. Soc.* **87**: 2020–2027.

Sarges, R., and Witkop, B. 1965c. Gramicidin. VII. The structure of valine- and isoleucine-gramicidin B. *J. Amer. Chem. Soc.* **87**: 2027–2030.

Schafer, J.A., Troutman, S.L., and Andreoli, T.E. 1974. Osmosis in cortical collecting tubules. ADH-independent osmotic flow rectification. *J. Gen. Physiol.* **64**: 228–240.

Schatzberg, P. 1965. Diffusion of water through hydrocarbon liquids. *J. Polymer Sci.* (No. 10, Pt. C): 87–92.

Schulman, J.H., and Teorell, T. 1938. On the boundary layer at membrane and monolayer interfaces. *Trans. Faraday Soc.* **34:** 1337–1342.

Sha'afi, R.I., and Feinstein, M.B. 1977. Membrane water channels and SH-groups. *Adv. Exp. Med. Biol.* **84:** 67–80.

Sha'afi, R.I., Gary-Bobo, C.M., and Solomon, A.K. 1971. Permeability of red cell membranes to small hydrophilic and lipophilic solutes. *J. Gen. Physiol.* **58:** 238–258.

Sha'afi, R.I., Rich, G.T., Sidel, V.W., Bossert, W., and Solomon, A.K. 1967. The effect of the unstirred layer on human red cell water permeability. *J. Gen. Physiol.* **50:** 1377–1399.

Sidel, V.W., and Solomon, A.K. 1957. Entrance of water into human red cells under an osmotic pressure gradient. *J. Gen. Physiol.* **41:** 243–257.

Singer, S.J., and Nicolson, G.L. 1972. The fluid mosaic model of the structure of cell membranes. *Science* **175:** 720–731.

Small, D.M. 1967. Phase equilibria and structure of dry and hydrated egg lecithin. *J. Lipid Res.* **8:** 551–557.

Solomon, A.K. 1968. Characterization of biological membranes by equivalent pores. *J. Gen. Physiol.* **51:** 335s–364s.

Solomon, A.K., Chasan, B., Dix, J.A., Lukacovic, M.F., Toon, M.R., and Verkman, A.S. 1983. The aqueous pore in the red cell membrane: Band 3 as a channel for anions, cations, nonelectrolytes, and water. *Ann. N.Y. Acad. Sci.* **414:** 97–124.

Spring, K.R. 1983. Fluid transport by gallbladder epithelium. *J. Exp. Biol.* **106:** 181–194.

Staverman, A.J. 1951. The theory of measurement of osmotic pressure. *Rec. Trav. Chim.* **70:** 344–352.

Steck, T.L. 1974. The organization of proteins in the human red blood cell membrane. A review. *J. Cell. Biol.* **62:** 1–19.

Stevens, C.F., and Tsien, R.W. 1979. *Membrane Transport Processes,* Vol. 3: *Ion Permeation Through Membrane Channels.* Raven Press, New York.

Suzuki, K., and Taniguchi, Y. 1972. Effect of pressure on biopolymers and model systems. In *Symposium of the Society for Experimental Biology,* No. 26: *The Effects of Pressure on Organisms,* Eds., M.A. Sleigh and A.G. MacDonald. Academic Press, New York, pp. 103–124.

BIBLIOGRAPHY

Tanford, C. 1961. *Physical Chemistry of Macromolecules.* John Wiley & Sons, New York.

Teorell, T. 1953. Transport processes and electrical phenomena in ionic membranes. *Progr. Biophys. Biophysic. Chem.* **3**: 305–369.

Terwilliger, T.C., and Solomon, A.K. 1981. Osmotic water permeability of human red cells. *J. Gen. Physiol.* **77**: 549–570.

Tredgold, R.H., and Jones, R. 1979. A study of gramicidin using deuterium oxide. *Biochim. Biophys. Acta* **550**: 543–545.

Urban, B.W., and Hladky, S.B. 1979. Ion transport in the simplest single file pore. *Biochim. Biophys. Acta* **544**: 410–429.

Urban, B.W., Hladky, S.B., and Haydon, D.A. 1980. Ion movement in gramicidin pores. An example of single-file transport. *Biochim. Biophys. Acta* **602**: 331–354.

Urry, D.W. 1971. The gramicidin A transmembrane channel: A proposed $\Pi_{(L,D)}$ helix. *Proc. Natl. Acad. Sci. USA* **68**: 672–676.

Urry, D.W. 1972. Protein conformation in biomembranes: Optical rotation and absorption of membrane suspensions. *Biochim. Biophys. Acta* **265**: 115–168.

Urry, D.W., Long, M.M., Jacobs, M., and Harris, R.D. 1975. Conformation and molecular mechanisms of carriers and channels. *Ann. N.Y. Acad. Sci.* **264**: 203–220.

Urry, D.W., Prasad, K.U., and Trapane, T.L. 1982a. Location of monovalent cation binding sites in the gramicidin channel. *Proc. Natl. Acad. Sci. USA* **79**: 390–394.

Urry, D.W., Walker, J.T., and Trapane, T.L. 1982b. Ion interactions in (1-^{13}C) D-Val[8] and D-Leu[14] analogs of gramicidin A, the helix sense of the channel and location of ion binding sites. *J. Membrane Biol.* **69**: 225–231.

Van't Hoff, J.H. 1887. Die Rolle des osmotischen Druckes in der Analogie zwischen Lösungen und Gasen. *Z. physik. Chemie* **1**: 481–493. A translation of an exerpt from this paper is given in *Benchmark Papers in Human Physiology,* Vol. 12: *Cell Membrane Permeability and Transport,* Ed., G.R. Kepner. Dowden, Hutchingson and Ross, Stroudsburg, Pa., 1979, pp. 20–28.

Varanda, W., and Finkelstein, A. 1980. Ion and nonelectrolyte permeability properties of channels formed in planar lipid bilayer membranes by the cytolytic toxin from the sea anemone, *Stoichactis helianthus. J. Membrane Biol.* **55**: 203–211.

Vegard, L. 1908. On the free pressure in osmosis. *Proc. Cambridge Phil. Soc.* **15**: 13–23.

Vieiera, F.L., Sha'afi, R.I., and Solomon, A.K. 1970. The state of water in human and dog red cell membranes. *J. Gen. Physiol.* **55**: 451–466.

Villegas, R., and Villegas, G.M. 1960. Characterization of the membranes in the giant nerve fibre of the squid. *J. Gen. Physiol.* **43** (Suppl. 1): 73–104.

Vreeman, H.J. 1966. Permeability of thin phospholipid films. III. Experimental method and results. *Kon Ned. Akad. Wetensch. Proc. Ser. B.* **69**: 564–577.

Wade, J.B. 1980. Hormonal modulation of epithelial structures. *Current Topics in Membranes and Transport* **13**: 123–147.

Walter, A., and Gutknecht, J. 1984. Monocarboxylic acid permeation through lipid bilayer membranes. *J. Membrane Biol.* **77**: 255–264.

Walter, A., and Gutknecht, J. 1986. Permeability of small nonelectrolytes through lipid bilayer membranes. *J. Membrane Biol.* **90**: 207–217.

Weinstein, S., Wallace B.A., Blout, E.R., Morrow, J.S., and Veatch, W. 1979. Conformation of gramicidin A channel in phospholipid vesicles: A ^{13}C and ^{19}F nuclear magnetic resonance study. *Proc. Natl. Acad. Sci. USA* **76**: 4230–4234.

Wheeler, T.J., and Hinkle, P.C. 1985. The glucose transporter of mammalian cells. *Ann. Rev. Physiol.* **47**: 503–517.

Whittembury, G., Carpi-Medina, P., Gonzalez, E., and Linares, H. 1984. Effect of para-chloromercuribenzenesulfonic acid and temperature on cell water osmotic permeability of proximal straight tubules. *Biochim. Biophys. Acta* **775**: 365–373.

Wieth, J.O., and Brahm, J. 1977. Separate pathways to water and urea in red blood cells? A comparative physiological approach. *Proc. Int. Congr. Physiol. Sci. 27th Paris,* vol. 12, p. 126 (Abstract 1.06).

Wright, E.M., and Pietras, R.J. 1974. Routes of nonelectrolyte permeation across epithelial membranes. *J. Membrane Biol.* **17**: 293–312.

Zimmerberg, J., and Parsegian, V.A. 1986. Polymer inaccessible volume changes during opening and closing of a voltage-dependent ionic channel. *Nature.* **323**: 36–39.

Index

INDEX

Diffusion coefficient, 13
  of potassium ions, 149
  of sodium ions, 149
  of water, 149
Diffusion permeability coefficient, 33
  effect of unstirred layers, 38
Diffusion theory (restricted), 115
Diffusive flow, 28
Diffusive flux term, 28
Diisothiocyano-2,2'-stilbene disulfonate
  (DIDS), 178, 179
Diols, permeability in red cell membrane, 172
Dithiobis-2-nitrobenzoic acid (DTNB), 178
Double file pore, 52

Electrokinetic effects, 139
Electroneutrality, 144
Electroosmosis, 55, 78, 139
  equivalence to streaming potentials, 58
  experiments, 56
  and hydrostatic pressure, 80
  measurement, 140
  through permselective pore, 56
Electrostatic energy barrier, 135, 148
Epithelial transport, 185 et seq.
Epithelia:
  antidiuretic hormone-induced permeability,
    186
  effect of antidiuretic hormone, 191 et seq.
  functions, 185
  leaky epithelia, 186
  mucosal side, 186
  paracellular route, 186
  passive ion permeability, 187
  serosal side, 186
  transcellular pathway, 186
  transcellular water flow, 188
  water permeability:
    of basolateral membrane, 188
    general characteristics, 186 et seq.
    of luminal membrane, 188
  water permeability coefficients, 188
Ergosterol, 107
Erythrocyte ghosts, 168
Erythrocyte membrane, see Red cell membrane
Ethanolamine group, 130
Ethylene glycol:
  reflection coefficient, 173
  permeability in red cell membrane, 172
  transport across red cell membrane, number
    of pores, 183
  transport inhibition by copper, 173
N-Ethylmaleimide, 178
Exocytosis, 93

Fick's law, 15, 18
Filipin, 107
  molecular structure, 108

Filtration permeability coefficient, 10
  effect of unstirred layers, 38
  temperature dependence, 103, 158
Fluid motion, equivalence of $\Delta\pi$ and ΔP, 12
Formamide, 100
Formic acid, 98
Formyl group, in gramicidin A, 130
Freeze etch experiments, 178
Frictional drag, 44
Frictional coefficient, 44, 87
Fundulus egg, water permeability coefficient,
  155

Gall bladder, 186
Gibbsian equilibrium condition, 4
Glucose, 108
  transport across red cell membrane, 172
Glucose-6-phosphate dehydrogenase, 108
Glycerol, transport in red cell membrane, 183
Glycerol monoolein membranes, 139
Glycophorin, 177
Gramicidin A, 31, 130 et seq.
  amino acid sequence, 130
  cation selective pore formation, in lipid
    bilayers, 130
  effect on glycerol monoolein membranes,
    139
  effect on lipid bilayers, 130 et seq.
    conductance, 134, 137
    ion-water interaction, 139 et seq.
    sodium transport, 135
  ethanolamine group, 130
  formyl group, 130
  pore formation in lipid bilayers:
    comparison with amphotericin B, 149
    comparison with nystatin, 149
    hydraulic permeability coefficient, 144
    hydrophobic surface, 132
    ion effect on water permeability, 143
    ion permeability, 134
    ion wall interactions, 145
    ion-water interaction, 139
    lithium ion binding, 146
    molecular model, 133
    polar groups, 132
    pore-forming molecules, 132
    pore model, 132
    pore structure, 132
    potassium ion binding, 146
    proton transport, 134
    single-file transport, 135
    sodium ion binding, 146
    streaming potentials, 140
    unidirectional ion flux, 146
    water molecules (number) in pore, 135, 140
    water permeability, 135
    water permeability per pore, 138
    water structure within pore, 142

224

water transport, 142
water-water interactions, 141
RNA polymerase inhibition, 130
Grotthus mechanism, 141

Heptaene, 109
Hexadecane, 95, 102
Hexanediol, 100, 162, 193
Hydraulic permeability coefficient, 10, 87
Hydrogen bonding, 142
Hydrostatic pressure, 8, 71
  and channel gating, 25
  and water flow, 12
Hydroxyl group, in amphotericin B, 117

Impermeant solute, 10
Interdiffusion, 83
Intracellular organelles, 93
Iodoacetamide, 178
Ionic physics, 140
Ions, single-file transport, 42
Ion transport, carrier mediated, 93
Ion-water interaction, in gramicidin A treated
  membranes, 139
Irreversible thermodynamics, 59, 84 et seq.
  and osmosis, 84
  phenomenological coefficients, 87
  reflection coefficient, 87
Isobutyramide, 100, 162, 193

Laminar flow, 28
Lipid bilayer membranes, 91 et seq.
  amphotericin B treated, solvent drag, 160
  area occupied by pores, 128
  butanol permeability, 106
  cholesterol-containing, 103
  conductance induced by gramicidin A, 131
  diffusion coefficients, 101
  diffusion permeability coefficient, 102
  effect of amphotericin B, 107 et seq.
    anion selectivity, 120
    cation permeability, 119
    pore formation, 110
    pore size, 114
  effect of gramicidin A, 130 et seq.
    cation permeability, 131
    conductance, 137
    ion permeability, 134
    ion-water interaction, 139
    proton transport, 134
    sodium transport, 135
    urea impermeability, 134
    water permeability, 135
    water permeability coefficient, 139
    water-water interaction, 141
  effect of nystatin:
    anion selectivity, 120
    cation permeability, 119

pore formation, 114
pore size, 114
filtration permeability coefficient, 103
fluidity, 98, 104
formation, 101
gramicidin A-treated:
  ion effect on water permeability, 143
  unidirectional ion flux, 146
hydrocarbon tails, 95
hydrophobic interior, 98
impermeability to ions, 94
ion permeability, effect on water
  permeability, 145
ion-wall interactions, 145
lipid chain length, 104
lipid chain unsaturation, 104
lipid composition, 100
membrane conductance, 109
nonelectrolyte permeability coefficients, 96
as oil membrane, 94
partition coefficient, 101
permeability coefficient:
  of formamide, 100
  of nonelectrolytes, 99
  of water, 99
permeability properties, 94
  quantitative considerations, 96
permeability to nonpolar lipophilic solutes,
  94
permeability to polar non-electrolytes, 94
permselective single file pores, 144
thickness, 120
unstirred layers, 105
viscosity, 106
water permeability per pore, 123
Lipids, polar head groups, 94
Lipophilic molecule, 105
Lipophilic solute, 64
Luminal membranes, 188

Membrane:
  permeability to urea, 81
  water permeability coefficients, 32
Membrane conductance, of gramicidin A
  treated membranes, 137
Membrane permeability, 3
Membrane physiology, 93
Membrane potential, 23
Membrane protein, 166
Membrane-solution interface, 12
Methylene bridge, 109
Micelles, 95
Molecular sieving, 110
Monoglyceride/ergosterol membranes, 121
Monoglycerides, 121
Mucosamine, 109

Negative osmosis, 64

INDEX

Polar groups, in gramicidin A induced pores, 132
Polar solute, in oil membrane, 65
Polyene chromophore, 117
Polyenes, 107
Polyhydroxylic lactone, 107
Pores, in red cell membrane, 176
Porous membrane, 18
  chemical potential of water, 20
  diffusion permeability coefficient, 35
    pore radius, 36
  ion selective, 56
  macroscopic pore:
    partition coefficient into pore, 70
    pressure within, 76
    solute chemical potential, 71
    solute distribution within pore, 73
    solute molecule size, 73
  osmosis:
    convective hydraulic flow, 71
    induced by permeant solute, 59
    macroscopic pore, 68
    pore radius, 35
    single-file pore, 42, 66
    solute molecule size, 73
    symmetric solutions, 76
  osmotic flow, 21
  permselective single-file pores, 56
  pore pressure gradient, 83
  pore radius, 27, 35
    calculation, 28
    solvent molecule radius, 29, 37
  pressure gradient, 20
  single-file pore, 42, 66
  solvent drag, 160
  water permeability calculations, 35
    laminar flow, 36
Potassium ions, diffusion coefficient, 149
Proton transport:
  in ADH-induced pores, 201
  in gramicidin A treated membranes, 134
  Grotthus mechanism, 141
  in red cell membrane, 176

Quasi-laminar flow, 156

Red cell membrane:
  acetamide permeability, 172
  amide permeability, 172
  anion exchange protein, 179
  diffusion permeability coefficient, 167
  diol permeability, 172
  ethylene glycol permeability, 172
  glucose transport, 172
  nonelectrolyte permeability, 172
  number of water pores, 177
  osmotic water entry, 167
  pathways for nonelectrolytes, 172
  pores, 166

  characteristics, 174
  for glycerol transport, 183
  ion conductance, 175
  ion impermeability, 175
  permeability properties, 175
  radius, 167, 174
  single-file mechanism, 175
  for urea permeability, 180
  proteins, 177
  saturation kinetics of urea, 180
  sulfhydral groups, 178
  tracer exchange experiments, 168
  unidirectional urea flux, 181
  urea permeability, 172
  water flow:
    through lipid bilayers, 169
    through pores, 169, 171
  water permeability coefficients, 167
  water permeation calculations, 170
Reflection coefficients, 60
  amphotericin B pore, 112
  irreversible thermodynamics, 84 et seq.
  nystatin pore, 112
  one-dimensional result, 80
  plasma membranes, 163
  red cell membrane, 172
  single-file pore, 68
  three-dimensional result, 80
Reynolds number, 27
RNA polymerase, inhibition, 130

Saturation kinetics, 180
Single channel conductance, 124, 151
Single-file transport, 42 et seq.
  continuum theory, 50
  diffusive flux, 49
  electroosmosis, 55
  equilibrium, 48
  in gramicidin A treated membranes, 135
  kinetics, 53
  pore length, 52
  solvent flow, 45, 51
  statistical mechanical calculation, 45
  tracer diffusion, 54
    of water, 46
  tracer flux-ratio, 53
  water permeability coefficient calculations, 42, 46, 49
Sodium channels, 23
  in squid axon and mode of Ranvier, 164
Sodium transport, in gramicidin-treated membranes, 135
Soil microorganisms, 130
Solubility diffusion mechanism, 12
Solute-solvent coupling, in osmosis, 84
Solvent drag, 85, 160
  experimental studies, 161
Sphingomyelin, 98
Sterols, 94, 107

227